T0325186

An Analysis of

Amartya Sen's

Development as Freedom

Janna Miletzki
with
Nick Broten

Published by Macat International Ltd
24:13 Coda Centre, 189 Munster Road, London SW6 6AW.

Distributed exclusively by Routledge
2 Park Square, Milton Park, Abingdon, Oxon OX14 4RN
711 Third Avenue, New York, NY 10017, USA

Routledge is an imprint of the Taylor & Francis Group, an informa business

www.macat.com
info@macat.com

Cataloguing in Publication Data
A catalogue record for this book is available from the British Library.
Library of Congress Cataloguing-in-Publication Data is available upon request.
Cover illustration: Etienne Gilfillan

ISBN 978-1-912302-39-0 (hardback)
ISBN 978-1-912127-04-7 (paperback)
ISBN 978-1-912281-27-5 (e-book)

CONTENTS

THE MACAT LIBRARY

The Macat Library is a series of unique academic explorations of seminal works in the humanities and social sciences – books and papers that have had a significant and widely recognised impact on their disciplines. It has been created to serve as much more than just a summary of what lies between the covers of a great book. It illuminates and explores the influences on, ideas of, and impact of that book. Our goal is to offer a learning resource that encourages critical thinking and fosters a better, deeper understanding of important ideas.

Each publication is divided into three Sections: Influences, Ideas, and Impact. Each Section has four Modules. These explore every important facet of the work, and the responses to it.

This Section-Module structure makes a Macat Library book easy to use, but it has another important feature. Because each Macat book is written to the same format, it is possible (and encouraged!) to cross-reference multiple Macat books along the same lines of inquiry or research. This allows the reader to open up interesting interdisciplinary pathways.

To further aid your reading, lists of glossary terms and people mentioned are included at the end of this book (these are indicated by an asterisk [*] throughout) – as well as a list of works cited.

Macat has worked with the University of Cambridge to identify the elements of critical thinking and understand the ways in which six different skills combine to enable effective thinking.
Three allow us to fully understand a problem; three more give us the tools to solve it. Together, these six skills make up the **PACIER** model of critical thinking. They are:

ANALYSIS – understanding how an argument is built
EVALUATION – exploring the strengths and weaknesses of an argument
INTERPRETATION – understanding issues of meaning

CREATIVE THINKING – coming up with new ideas and fresh connections
PROBLEM-SOLVING – producing strong solutions
REASONING – creating strong arguments

To find out more, visit **WWW.MACAT.COM.**

CRITICAL THINKING AND *DEVELOPMENT AS FREEDOM*

Primary critical thinking skill: EVALUATION
Secondary critical thinking skill: REASONING

Amartya Sen uses his 1999 work *Development as Freedom* to evaluate the processes and outcomes of economic development.

Having come to the conclusion that development is best summed up as the expansion of freedom, Sen examines traditional definitions and understandings of the term. He says people tend to think of freedoms as economic (the freedom to enter into market exchanges) or political (the freedom to vote and be an active citizen), and tries to understand why the definition has been so narrow hitherto. He concludes that an evaluation of true freedom must necessarily include the freedom to access social services such as healthcare, sanitation and nutrition, just as much as it must acknowledge economic and political freedoms.

Evaluating the relevance of the current thinking behind development, Sen's concludes that the term 'freedom' cannot simply be about income. In many ways, measuring income does not account for various "unfreedoms" (manmade or natural bars to wellbeing) that hinder development. Sen's evaluation is all the more powerful for its clarity: "The freedom-centered perspective has a generic similarity to the common concern with 'quality of life.'"

ABOUT THE AUTHOR OF THE ORIGINAL WORK

Amartya Sen is one of the most influential economists of the twentieth century. A Nobel Prize winner in economics and a senior fellow of the Harvard Society of Fellows, he has produced work characterized by cultural breadth and diversity.
Born in India in 1933 to a prominent family of scholars and academics, Sen studied first at Calcutta University, then at Cambridge. He has lived and worked in India, Britain, and the United States, and his holistic approach to understanding what "development" actually means reflects this broad life experience. Sen draws from a rich well of thinking to approach the issue, investigating philosophy, formal economics, general culture, social and economic facts, and even anecdotal tales to arrive at innovative conclusions.

ABOUT THE AUTHORS OF THE ANALYSIS

Janna Miletzki is researching for a PhD in human geography at LSE.

Nick Broten was educated at the California Institute of Technology and the London School of Economics. He is doing postgraduate work at the Pardee RAND Graduate School and works as an assistant policy analyst at RAND. His current policy interests include designing distribution methods for end-of-life care, closing labour market skill gaps, and understanding biases in risk-taking by venture capitalists.

ABOUT MACAT

GREAT WORKS FOR CRITICAL THINKING

Macat is focused on making the ideas of the world's great thinkers accessible and comprehensible to everybody, everywhere, in ways that promote the development of enhanced critical thinking skills.

It works with leading academics from the world's top universities to produce new analyses that focus on the ideas and the impact of the most influential works ever written across a wide variety of academic disciplines. Each of the works that sit at the heart of its growing library is an enduring example of great thinking. But by setting them in context – and looking at the influences that shaped their authors, as well as the responses they provoked – Macat encourages readers to look at these classics and game-changers with fresh eyes. Readers learn to think, engage and challenge their ideas, rather than simply accepting them.

'Macat offers an amazing first-of-its-kind tool for interdisciplinary learning and research. Its focus on works that transformed their disciplines and its rigorous approach, drawing on the world's leading experts and educational institutions, opens up a world-class education to anyone.'

Andreas Schleicher
Director for Education and Skills, Organisation for Economic Co-operation and Development

'Macat is taking on some of the major challenges in university education ... They have drawn together a strong team of active academics who are producing teaching materials that are novel in the breadth of their approach.'

Prof Lord Broers,
former Vice-Chancellor of the University of Cambridge

'The Macat vision is exceptionally exciting. It focuses upon new modes of learning which analyse and explain seminal texts which have profoundly influenced world thinking and so social and economic development. It promotes the kind of critical thinking which is essential for any society and economy. This is the learning of the future.'

Rt Hon Charles Clarke, former UK Secretary of State for Education

'The Macat analyses provide immediate access to the critical conversation surrounding the books that have shaped their respective discipline, which will make them an invaluable resource to all of those, students and teachers, working in the field.'

Professor William Tronzo, University of California at San Diego

WAYS IN TO THE TEXT

KEY POINTS

- Born in 1933, Amartya Sen is an Indian economist who has made major contributions to development economics* (a subfield of economics describing how premodern economies transition to modern economies, alleviating poverty) and welfare economics* (a subfield concerned with how economic well-being is created and distributed throughout society).
- Published in 1999, *Development as Freedom* gives a many-sided approach to development* (the process by which a low-income nation increases its prosperity and the well-being of its citizens by following polices designed to promote structural change) based on the idea that successful development both requires and leads to the expansion of human freedom.
- The book offers a perspective on development that falls between liberalization* (approaches in which privatization is encouraged and the markets* are allowed to operate unhindered by government intervention) and state-based* approaches (according to which the government intervenes in the functioning of an economy, to ensure a measure of social justice, for example).

Who Is Amartya Sen?

Amartya Sen, the author of *Development as Freedom* (1999), is one of the most influential economists of the twentieth century. He currently teaches at Harvard University and until 2004 was the master of Trinity College at the University of Cambridge. He has taught economics at many other institutions around the world and is a senior fellow at the Harvard Society of Fellows,* an important academic exchange based at Harvard covering multiple disciplines. These highly respected appointments show the importance of Sen's work. Indeed, he was awarded the Nobel Prize in Economics* in 1998 for his contributions to welfare economics. By his own admission, Sen has held no significant nonacademic job.

Sen was born in India in 1933 to a prominent family of scholars and academics. He has lived and worked in India, Britain, and the United States, and his academic approach reflects this diversity. *Development as Freedom* can be seen as the work of a scholar looking for common ground between disciplines and perspectives.

Though the book is largely descriptive, much of Sen's work is highly rigorous, applying mathematical reasoning to complex social problems. He began his academic career in the subfield of economics known as social choice theory,* the aim of which is to understand the limitations of various procedures for aggregating (fully quantifying) information in a group. There is an interesting parallel between that work and Sen's later contributions to development economics because both deal with how societies and individuals relate. Although Sen's approach to development is focused on the expansion of individual freedoms, *Development as Freedom* is also a study of social institutions and the outcomes they produce.

What Does *Development as Freedom* Say?

Development as Freedom presents a framework for understanding the processes and outcomes of economic development. Fundamentally, the book argues that development is the expansion of freedom and the elimination of what Sen calls "unfreedoms":* man-made or natural bars to well-being, of which a lack of access to health care, sanitation, nutrition due to famine, or basic political and civil rights are classic examples. For Sen, "freedom" can include, among other things, economic freedom, political freedom, and the freedom to access social goods and services: put simply, the ability to live a life worth living.

Sen's broad definition of freedom is in part what makes the book unique. When people who subscribe to the political philosophy of libertarianism* discuss freedom, for example, they are likely referring to specific economic rights such as the ability to enter into market exchanges* (the interactions between buyers and sellers in a market). In the political world, "freedom" often refers to citizenship rights such as suffrage (the right to vote). Sen's approach is much broader. The right to enter into exchange is included in his list of freedoms, but he notes that lack of access to other freedoms such as education or health care may restrict economic freedoms for some people—even if the laws of a society are unrestricted. He also says that development cannot just be about income: in many cases measuring income does not account for various "unfreedoms" that block development. The simplicity of his basic premise is clear: "the freedom-centered perspective has a generic similarity to the common concern with 'quality of life.'"[1]

At the heart of Sen's freedom approach is the concept of "capabilities,"* or the capacity to perform the functions that comprise a good life. Sen suggests that the best way to compare and evaluate the freedoms available to different people in different societies is to examine their "capability set"—in other words, the maximum number of life-fulfilling functions available to that particular group.[2] For example, in a comparison of college students and the working poor in the United States, Sen might note that the capability set of many college students is larger than that of the working poor despite the two groups having very similar incomes. Students have access to education, and perhaps family support, and can also use their time to develop human capital (the intellectual, physical, and personal capacities of people, considered to be an important part of economic expansion). All these freedoms are less likely to be available to the working poor.

Sen applies his freedom approach to several controversial topics in development economics. Perhaps most importantly, he discusses the appropriate role of markets and states in economic development, concluding that both are important for expanding freedom. He also argues that democratic control is crucial for development on the grounds it often leads to other freedoms necessary for this to take place. Sen also discusses the importance of women's rights, and suggests that discrimination against women has a damaging effect on all people in society. Another important topic for Sen is culture: he dismisses the idea that some cultures are more favorable to economic development than others.

Why Does *Development as Freedom* Matter?

Development as Freedom is a key work in the field of development economics. It brings together many views on development into a single arena that is focused on the idea of freedom. In this way, the book appeals to those who support market-based approaches to economic development—the policies known as the Washington Consensus,* for example, designed to make markets function more efficiently through things such as spending cuts—while also admitting that some market institutions can inhibit freedom for the poor. Sen's perspective on development is in some ways a middle path between market liberalization and state-based approaches, making the book appeal to readers with vastly differing viewpoints.

It is important to note, however, that Sen tries to rise above this conflict, not just resolve it. Those who believe in the power of the market will have to accept its limitations in terms of freedom: for example, people who cannot afford education or health care have far less chance to develop marketable skills. And those who question markets will have to understand that for market economies to function, people need to be able to enter into contracts, work, and perform other tasks.

This book will be useful for readers from many disciplines, particularly those with an interest in economics, development, or government, or those hoping to work in the nongovernmental sector. Reducing poverty is an important challenge for everyone, and readers of this book will be exposed to a well-balanced and innovative set of ideas that put forward ways of doing so.

The book may have the most direct impact on people studying economics. Sen is a broad thinker whose ideas regularly move from the economic to the philosophical. Students who are interested in economics, but find much of the work dry, may enjoy *Development as Freedom* because it directly addresses the most important questions in economic thought without becoming too mathematical. Students may also find that Sen's philosophy helps to make economics courses clearer by giving meaning to symbols used in economic models. For example, his discussion of the nature of well-being and utility* (an abstract measure of well-being used by economists to compare the outcomes of policies or institutional designs) is a powerful addition to the social choice theory* taught in introductory microeconomics. While the model describes how utility can be maximized, Sen asks how doing so fits with leading a meaningful life.

NOTES

1 Amartya Sen, *Development as Freedom* (New York: Anchor Books, 1999), 24.

2 Sen, *Development as Freedom*, 76.

SECTION 1
INFLUENCES

MODULE 1
THE AUTHOR AND THE HISTORICAL CONTEXT

KEY POINTS

- *Development as Freedom* presents a unique and rich framework for understanding development*—the process by which a low-income nation increases its prosperity and the well-being of its citizens by following policies designed to promote structural change.
- Sen was raised in an academic family and has spent his academic career at some of the world's highest-ranking universities.
- It is likely that Sen's background in a scholarly Brahmin* family inspired his later work on development. Brahmins are the highest caste—roughly, social class—in traditional Indian culture.

Why Read This Text?

Development as Freedom (1999) is the culmination of the economist Amartya Sen's work on economic development. It represents the ideas of a mature and experienced thinker with life experience on three continents. Prior to *Development as Freedom* much academic work on development tended to emphasize either the importance of the market* (the institution in which willing sellers of goods and willing buyers conduct their business) or the role of the state* (that is, the government) in creating and regulating markets. But this book develops a unified approach. By focusing on freedom, and specifically the capabilities* required to function with freedom in society ("capabilities" being the capacity of individuals to do the things that

> **“This work outlines the need for an integrated analysis of economic, social and political activities, involving a variety of institutions and many interactive agencies. It concentrates particularly on the roles and interconnections between certain crucial instrumental freedoms, including *economic opportunities, political freedoms, social facilities, transparency guarantees*, and *protective security.*”**
>
> Amartya Sen, *Development as Freedom*

make their lives meaningful), Sen bridges the intellectual gap between these two perspectives—simply put, markets are important for creating an environment of economic freedom but are by no means sufficient for creating genuine capability.

Anyone interested in economic development will find *Development as Freedom* is not only a useful guide through the many different perspectives on development but also a persuasive account of Sen's own view. In the book's 298 pages, Sen discusses the ancient Greek philosopher Aristotle's* views on the meaning of economic life, the eighteenth-century thinker Adam Smith's* rules for economic growth, the moral philosopher John Rawls's* theory of justice, and the political philosopher Robert Nozick's* interpretation of libertarianism,* according to which governmental intervention in the life of the citizen is interpreted as an undesirable challenge to liberty, among other perspectives.

The book is not just a unique example of original economic analysis. It also forces readers committed to any one perspective to consider the strengths and weaknesses of other views. Sen is more philosophical than many economists, and his willingness to apply rigorous analysis to such broad issues as human freedom is one of the reasons his work has endured.

Author's Life

Amartya Sen is perhaps best known as the winner of the Nobel Prize in Economics* in 1998. The award brought him a much wider audience and may be responsible for the prominence of *Development as Freedom*, published one year later. Sen was born in the state of West Bengal, India, in 1933. His parents were both from Dhaka, the capital of modern-day Bangladesh, where his father was a university professor of chemistry. Perhaps inspired by this academic background, Sen studied economics at the University of Calcutta and later at Trinity College in Cambridge, where he received his PhD. He has taught at several prominent universities, including the Massachusetts Institute of Technology, the University of California at Berkeley, Stanford University, Cornell University, the University of Calcutta, the Delhi School of Economics, the London School of Economics, the University of Oxford, and Harvard University.

Development as Freedom reflects Sen's own experiences living in India, Great Britain, and the United States. Examples used in the book not only show the contrast between rich and poor people within developing countries but also the inequalities that exist within the developed world. For example, he describes a case of political violence he witnessed as a child, saying: "It made me reflect, later on, on the terrible burden of narrowly defined identities, including those firmly based on communities and groups."[1] It is likely that Sen's holistic approach to development (that is, an approach that considers the integration of development's many constituent features) was influenced by this broad life experience. Another important feature of Sen's personal life is that he rarely had to deal with censorship or restrictions on his writing: "Since I have been fortunate in living in three democracies with largely unimpeded media (India, Britain, the United States), I have not had reason to complain about any lack or opportunity of public presentation."[2]

Author's Background

Amartya Sen has lived a long and active life, and his career has coincided with a number of monumental events, including the tense global standoff between the United States and the Soviet Union and the nations aligned to them known as the Cold War* (roughly 1946–91), the expansion of global markets, and the rise of political self-determination in many former colonies. Yet much of his work seems to stem from older philosophical traditions, and his blending of multiple traditions is notable.

Sen was born into a Brahmin* family (that is, a family belonging to the highest caste in India's traditional system of social stratification), a legacy that is apparent in his work. Beyond Sen's rigor as an economist, his cultural breadth may be his most significant characteristic as a scholar. He fleshes out his arguments with anecdotes and socioeconomic facts from India, as well as with excerpts from Indian philosophical texts. His writings on famine, for instance, were inspired by his childhood experience of the Bengal famine of 1943* (a catastrophe that led to the deaths of around three million people and significant social disruption in India). Further, his discussions of culture draw from a deep historical knowledge, combined with lived experience.

It is likely that Sen draws on his own family's traditions—his maternal grandfather Kshiti Mohan Sen was a scholar of medieval Indian literature and taught the ancient scriptural language of Sanskrit at the University of Santiniketan.[3] Still, his perspective remains broad. For example, in a discussion of the perception that Asian culture is more totalitarian than the Western tradition (the perception that it is a culture in which the individual's life is subject to a degree of regulation incompatible with personal liberty), he writes: "It is by no means clear to me that [the Chinese philosopher] Confucius* is more authoritarian in [terms of order and discipline] than, say, [the ancient Greek philosopher] Plato* or [the early Christian thinker] St. Augustine."*[4]

NOTES

1 Amartya Sen, *Development as Freedom* (New York: Anchor Books, 1999), 8.

2 Sen, *Development as Freedom*, xiv.

3 Amartya Sen, "Amartya Sen—Biographical," Nobelprize.org, accessed October 14, 2015, http://www.nobelprize.org/nobel_prizes/economic-sciences/laureates/1998/sen-bio.html.

4 Sen, *Development as Freedom*, 234.

MODULE 2
ACADEMIC CONTEXT

KEY POINTS

- **The field of economics is concerned with how wealth is created and distributed in society.**
- **Sen's career has coincided with the development of mathematical economics and important debates about the proper role of the government in the economy.**
- **Two important influences on Sen are the economist Kenneth Arrow* and the moral philosopher John Rawls,* both of the United States.**

The Work in Its Context

Amartya Sen's *Development as Freedom* is a work of economics. As such, it is part of an academic tradition mainly concerned with how goods and services are produced and distributed in society. More specifically, a major drive of economic research deals with the origins of wealth.* In other words, it asks the question: Why are some countries prosperous while others are not?

The so-called father of modern economics is the Scottish thinker Adam Smith,* whose book *The Wealth of Nations* (1776) was one of the first to address this question rigorously. For Smith, allowing markets*—the institutions that facilitate trade—to operate freely and openly was essential for creating wealth. He believed humans had a natural tendency to "truck, barter, and exchange," and that if individuals pursued their own interests in a market context, society as a whole would be served.[1]

While Smith's faith in market systems to achieve social goals effectively has possibly been overstated by some of his followers, his name is still often associated with support for free markets—a view

> **“Economics is the science which studies human behavior as a relationship between ends and scarce means which have alternative uses.”**
>
> Lionel Robbins, *An Essay on the Nature and Significance of Economic Science*

that many economists have challenged. Perhaps the loudest of voices critiquing the economic system of capitalism,* in which trade and industry are held in private hands and conducted for the sake of private profit, is the German economist and political philosopher Karl Marx,* whose book *Capital* (1867–94) presented a two-pronged critique of capitalism.[2] First, Marx suggested that the capitalist system would lead to the accumulation of wealth in fewer and fewer hands, eventually collapsing on itself. Second, Marx believed that companies would continue to exploit their employees to increase profits, which would eventually lead to the ruin of the worker. It is important to note that Marx saw himself as a scientist, merely describing historical patterns: he did not argue for improving welfare within capitalism. But Marx's criticism of capitalism did inspire others, such as the Fabians* in Britain, a social organization dedicated to increasing equality and opportunity, to stand up for working people within the capitalist system.

While literature dealing with economics is rich and diverse, the differences between Smith's and Marx's work show a central tension in the discipline. Economists are generally favorable toward markets and the economic dynamism that accompanies them, but the question of how best to care for the less fortunate remains.

Overview of the Field

Sen's intellectual career has intersected with several developments in economic thought. One such development was the introduction of formal mathematical tools to economics by such scholars as Kenneth Arrow,* the French economist Gérard Debreu,* and others, who were mainly concerned with elegant models of complete economies. An important feature of many of these mathematical models is that they describe an ideal state of the world in which the mathematical assumptions underlying them hold true. This work created an opportunity for scholars like Sen to build on the models so that they reflected the real world more accurately.

Beyond methodological developments, Sen's career has coincided with vehement discussions about the proper role of markets and states. Sen says in his Nobel Prize autobiography that when he studied at the University of Cambridge it was a battleground for these issues. Loosely speaking, the debate was between Keynesian* economists, named after the British economist John Maynard Keynes,* who promoted a stronger governmental role in the economy, and neoclassicists,* who were skeptical of such interventions. Sen recalls: "Even though there were a number of fine teachers who did not get very involved in these intense fights between different schools of thought … the political lines were, in general, very firmly—and rather bizarrely—drawn."[3] Sen's reluctance to join in on these discussions is a foretaste of his later work in *Development as Freedom*, which looks for a middle ground between the two perspectives.

Academic Influences

Sen is a scholar who draws on many influences—economics, philosophy, and even ancient Indian thinking.

Sen's most influential intellectual mentor was arguably Kenneth Arrow, whose impossibility theorem* inspired Sen to enter the field of social choice theory.* The impossibility theorem—describing, very roughly, how the rules of the various methods we use to make choices at levels as large as an electorate translate individual beliefs to rational group action—is derived from the field of social choice theory. While

Sen clearly respects Arrow's work, he often uses Arrow's results as a starting point for a broader critique. Much of Sen's social choice work, for example, is about reframing Arrow's original problem to show why the result is less pessimistic than it seems to be.

For example, Sen cites the proof that Arrow and the economist Gérard Debreu put forward that, given certain assumptions, "the results of the market mechanism are not improvable in ways that would enhance everyone's utility"*[4] ("utility" being a figurative measure of well-being used by social scientists to compare the outcomes of policies or institutional designs). In other words—markets are efficient, and when the market allocates goods and resources efficiently in society, no person can be made better off without others being made worse off. Sen uses this as a starting point for a broader discussion, however. He asks whether "the efficiency sought should not be accounted in terms of *individual freedoms*, rather than *utilities*."[5]

Another important influence on Sen's thought is John Rawls,* with whom Sen taught a course on social justice at Harvard.[6] Although Sen shows tremendous respect for Rawls, he challenges him in important ways. As Sen writes in his Nobel Prize autobiography: "If my work in social choice theory was initially motivated by a desire to overcome Arrow's pessimistic picture by going beyond his limited informational base, my work on social justice based on individual freedoms and capabilities* was similarly motivated by an aspiration to learn from, but go beyond, John Rawls's elegant theory of justice, through a broader use of available information."[7]

One aspect of Rawls's theory that Sen attempted to develop further was the "priority of liberty"—the idea that individual liberty should be considered more important than other considerations such as well-being. Sen's capabilities approach is a challenge to this idea.[8]

NOTES

1 Adam Smith, *An Inquiry into the Nature and Causes of the Wealth of Nations,* Library of Economics and Liberty, Book I, Chapter II, accessed September 13, 2015, http://www.econlib.org/library/Smith/smWN1.html#nn41.

2 Karl Marx, *Capital*, Library of Economics and Liberty, accessed September 13, 2015, http://www.econlib.org/library/YPDBooks/Marx/mrxCpContents.html.

3 Amartya Sen, "Amartya Sen—Biographical," Nobelpize.org, accessed October 14, 2015, http://www.nobelprize.org/nobel_prizes/economic-sciences/laureates/1998/sen-bio.html.

4 Amartya Sen, *Development as Freedom* (New York: Anchor Books, 1999), 117.

5 Sen, *Development as Freedom*, 117.

6 Sen, "Amartya Sen—Biographical."

7 Sen, "Amartya Sen—Biographical."

8 Sen, *Development as Freedom*, 63.

MODULE 3
THE PROBLEM

KEY POINTS

- *Development as Freedom* asks the question: What is economic development,* and how can it be achieved?
- Sen's views on development were presented in the context of the Washington Consensus* and the post-Washington Consensus perspectives on development. This is a perspective on economic development revolving around 10 specific policies, including reforms such as currency devaluation* and spending cuts, designed to make markets* function more efficiently.
- *Development as Freedom* was shaped by Sen's unique relationship with the World Bank,* an organization founded in 1944 with the explicit goal of reducing global poverty and which has played a central role in implementing economic reforms in developing countries.

Core Question

The core question that underlies Sen's analysis in *Development as Freedom* is: What is the meaning of economic development, and how can it best be achieved on a global scale?

The question of what development is and how it can be achieved is a fundamental one. It also suggests other questions: What do people actually value in their lives? How can they attain the things they value? These questions are not new. In the eighth century B.C.E., as Sen writes, texts in the ancient Indian language of Sanskrit were discussing the meaning of a successful life, noting "there is no hope of immortality by wealth."[1] Sen also draws on the ancient Greek philosopher Aristotle* to support his argument that life is about more than the accumulation of wealth. This is important in the context of the larger

> **“ Wealth is evidently not the good we are seeking; for it is merely useful and for the sake of everything else. ”**
> Aristotle, *Nicomachean Ethics*

discipline of economics. Most economic arguments treat money and utility* as identical or nearly identical, and equate material well-being with overall well-being. What is original about Sen's description of economic development is his shift of the traditional focus of economics into the human realm.

The question of what development actually *means* also has important policy implications. If development is seen simply as the raising of income, then the aim of policies should be to increase gross domestic product per capita (GDP is a measure of a country's economic output, stated as the value in US dollars of the goods and services, usually calculated annually). If the true meaning of development is broader, however, including such elements as education and health, then development policy should incorporate those aspects.

The Participants

Development scholars writing before the publication of *Development as Freedom* can be loosely grouped into two categories: those who saw market activities as the main path to development and those who argued for a wider role for the state* and other institutions. Supporters of the former view were often associated with the so-called Washington Consensus,* a set of ideas about desirable development policies proposed by the economist John Williamson.* The Washington Consensus reforms included:

- “fiscal discipline”—in other words, curbing government spending
- shifting spending priorities to pro-growth policies
- tax reform

- liberalizing interest rates—in other words, making markets work more freely
- currency devaluation* (a policy intended to make exports more affordable for the nation's trading partners)
- trade liberalization*
- liberalization of foreign investment
- privatization of industry
- deregulation—that is, curbing government influence on the workings of the economy and markets
- development and enforcement of property rights.[2]

Generally speaking, the Washington Consensus called for significant changes to countries' economies according to free-market principles, notably liberalization (the process of removing restrictions on market exchange, such as tariffs on trade between countries).

When this book was written, the neoliberal* agenda of the Washington Consensus emphasizing "liberalization, stabilization, and privatization"[3] gave way to a post-Washington Consensus[4] under the guidance of the director of the World Bank,* James Wolfensohn.* Rather than stressing market liberalization, this perspective focused on participation, good governance, and reducing poverty. Although Sen's book fits with this view in its effort to find a middle way between approaches to development favoring either markets or the state, it goes further by emphasizing that achieving development goals requires the removal of unfreedoms*—a lack of access to health care, sanitation, nutrition due to famine, or basic political and civil rights are good examples—that prevent people from living the lives they want to live.

The Contemporary Debate

While Sen has clearly remained engaged with the academic debate on development throughout his career, his ideas seem to be specifically "his." Nevertheless, the World Bank's shift from the neoliberal

Washington Consensus to Wolfensohn's more inclusive policies allowed Sen to present his ideas on development to a policy audience. Indeed, it was Wolfensohn who invited Sen to the World Bank to deliver the series of lectures that would later comprise *Development as Freedom*.

Sen's relationship with the World Bank, perhaps the leading global institution dedicated to economic development, is important. In the book, he writes: "The World Bank has not invariably been my favorite organization. The power to do good goes almost always with the possibility to do the opposite, and as a professional economist, I have had occasions in the past to wonder whether the Bank could not have done much better."[5] Still, Sen's willingness to speak to the Bank and his relationship with its president at the time indicates that he viewed the institution as moving in the right direction by taking up the post-Washington Consensus. This relationship also reveals Sen's position in the development conversation as an instigator rather than a radical outsider. Sen's own ideas do not fit neatly into one ideological category, and he has managed to work with members from various perspectives throughout his career.

NOTES

1 Amartya Sen, *Development as Freedom* (New York: Anchor Books, 1999), 13.

2 John Williamson, "A Short History of the Washington Consensus" (paper commissioned by Fundación CIDOB for a conference "From the Washington Consensus towards a new Global Governance," Barcelona, September 24–25, 2004, 3-4).

3 Dani Rodrick, "Goodbye Washington Consensus, Hello Washington Confusion? A Review of the World Bank's Economic Growth in the 1990s: Learning from a Decade of Reform," *Journal of Economic Literature* 44, no. 4 (2006): 973.

4 Joseph E. Stiglitz, *More Instruments and Broader Goals: Moving toward the Post-Washington Consensus* (Helsinki: UNU/WIDER, 1998).

5 Sen, *Development as Freedom*, XIII.

MODULE 4
THE AUTHOR'S CONTRIBUTION

KEY POINTS

- **Sen's goal in writing *Development as Freedom* was to change the conversation on economic development* by highlighting the importance of human freedom.**
- ***Development as Freedom* invented a new perspective on development that combined aspects of several competing theories.**
- ***Development as Freedom* is a unique and original contribution to the field of development economics.**

Author's Aims

Amartya Sen's main aim in writing *Development as Freedom* was to present a work on "development and the practical reasons underlying it, aiming at public discussion."[1] He wanted to advance the theory of development while attracting a general audience; even though the book is based on lectures given to the World Bank,* Sen aimed at a wider readership than economists and development practitioners. Further, he notes the work is "presented mainly for open deliberation and critical scrutiny."[2]

Sen rearranged the six original lectures to the Bank into 12 chapters, "for clarity and to make the written version more accessible to non-specialist readers."[3] The discussion—about development's ends and means—is "as nontechnical as possible."[4] Sen does not aim it at any particular interest group or government; rather, he attempts to reach the largest audience possible.

Despite this stated interest in a broader public debate and criticism, it would be wrong to assume that Sen was just playing the role of

> **" It is the customary fate of new truths to begin as heresies and end as superstitions. "**
>
> T. H. Huxley, *Science and Culture and Other Essays*

spokesman in writing this book. His ideas emerged from years of careful study in philosophy and economics and represent a deep understanding of the challenges of development. Sen's goal was to change the conversation, not just to amplify one aspect of it.

Approach

The basic proposition of *Development as Freedom* is that freedom is the building block of development. Sen's innovation was to present development as a process of expanding human freedom rather than simply increasing economic growth. He then added real substance to his message by developing the idea of capabilities*—the capacity of individuals to do the things that make their lives meaningful. As Sen writes: "Focusing on human freedoms contrasts with narrower views of development, such as identifying development with the growth of gross national product, or with the rise in personal incomes, or with industrialization, or with technological advance, or with social modernization."[5]

Sen's ambition is clear—his perspective, founded on freedom, is an attempt at a unified theory of development.

While this simple shift in perspective may seem obvious, compared with previous work on development it represents a profound transformation. Through the vehicle of freedom, Sen found a way to include many of the concerns of those opposed to the free market—trade unregulated by government intervention—into a view that recognizes the importance of markets and the economic freedom they represent.

Sen draws on his personal experience to develop and express this view. For example, his contribution to the study of famines as part of his capabilities approach stems from his own personal experience of the Bengal famine of 1943,* a catastrophe in which some three million people died. He is also inspired by his experience of living in India; the southern Indian state of Kerala* is frequently cited as an example of a place where income and capability are largely unrelated. In fact, the citizens of Kerala are much richer in terms of capability than their incomes would suggest. Without life experience in India, it is unlikely Sen would have been able to understand the politics of the region deeply enough to highlight the example of Kerala.

Contribution in Context

Though some aspects of Sen's analysis are not original, his freedom-based approach was highly original. A review of the book in the journal *Business Week* called it a "new approach" to combating "poverty and famine," and in many ways that assessment is correct.[6] Sen did not contribute an original technology to combat famine or extreme poverty, but his reframing of the problem revealed opportunities for solutions that had not been seen before.

That said, the seeds of the human development approach were planted before Sen made his contribution. In 1994—before *Development as Freedom* but after other work Sen had addressed to the topic—the economist T. N. Srinivasan* published the essay "Human Development: A New Paradigm or Reinvention of the Wheel," in which he suggested that the idea of measuring human development in ways other than income had appeared as early as the 1950s and was not unique to Sen.[7] But Sen's original contribution was not so much the idea that development is more than income, but the use of the concept of freedom to frame the discussion. Sen's original contribution is more philosophical than economic.

NOTES

1 Amartya Sen, *Development as Freedom* (New York: Anchor Books, 1999), xiii.

2 Sen, *Development as Freedom*, xiv.

3 Sen, *Development as Freedom*, xiii.

4 Sen, *Development as Freedom*, xiii.

5 Sen, *Development as Freedom*, 4.

6 Geri Smith, "Do Literacy and Health Spark Growth?" *Business Week*, September 20, 1999, accessed September 30, 2015, http://www.businessweek.com/1999/99_38/b3647034.htm.

7 Thirukodikaval Nilakanta Srinivasan, "Human Development: A New Paradigm or Reinvention of the Wheel?" *The American Economic Review* 84, no. 2 (1994): 238–43.

SECTION 2
IDEAS

MODULE 5
MAIN IDEAS

KEY POINTS

- **For Sen, freedom is both the means and the desired end of the development* process.**
- **Sen's argument is about expanding capabilities*—the capacity to perform the core functions of life.**
- ***Development as Freedom* is written in direct and plain language, and is intended for a general audience.**

Key Themes

The core idea in Amartya Sen's *Development as Freedom* is that economic development should be seen "as a process of expanding the real freedoms that people enjoy."[1] For Sen, the expansion of freedom is not only the desired *end* of a development process, but also the main *means* of development. In other words, any proposal for economic development should be judged by whether it expands the specific freedoms Sen considers essential. Among these are "economic opportunities, political freedoms, social facilities, transparency guarantees, and protective security."[2]

Acting against these essential freedoms are "unfreedoms."* Sen often treats the expansion of freedom and the elimination of unfreedom as identical. To Sen, unfreedoms can be man-made or natural. For example, he sees lack of nutrition due to famine as a crucial unfreedom, noting that political freedoms and food security* are connected: "It is not surprising that no famine has ever taken place in the history of the world in a functioning democracy."[3] In countries with elected leaders, governments will never allow the conditions to exist that might cause a famine because it would not be in their best

> **“ Development requires the removal of major sources of unfreedom: poverty as well as tyranny, poor economic opportunities as well as systematic social deprivation, neglect of public facilities as well as intolerance or overactivity of repressive states. ”**
>
> Amartya Sen, *Development as Freedom*

interest. Other unfreedoms include not having access to health care, sanitation, or basic political and civil rights.

Sen makes a distinction between what he calls the “process aspect of freedom”* (the capacity to perform certain aspects of freedom within the rules of a society) and the “opportunity aspect of freedom”* (the capacity actually to achieve freedoms in society).[4] While this distinction is somewhat academic, it allows what Sen hopes is a sufficiently broad definition of freedom. For example, while someone who subscribes to libertarian* ideas may favor unimpeded economic *rights* (the process aspect), Sen says it is also important to see whether all members of society fulfill their economic opportunities.

Exploring the Ideas

A critical component of Sen’s development framework is the idea of “capabilities,” specifically the capacity of “persons to lead the kind of lives they value—and have reason to value.”[5] Using the term “functionings”* to describe “the various things a person may value doing or being,” Sen defines a person’s capabilities as “the alternative combinations of functionings that are feasible for her to achieve.”[6] Consider, for example, a rich man who is fasting and a poor man who is starving. In their actual experience, in this case measured by consumption of food, their conditions are very similar. But the “capability set” of the rich man far exceeds that of the poor man. If he wanted to eat, he could, while the poor man could not.

To Sen, poverty is a form of capability deprivation. Two aspects of his discussion of poverty are particularly important. First, he argues that insufficient income is not the only contributing factor to capability deprivation. Second, the relationship between capabilities and income varies from country to country. Sen notes, for example, that African American men have a much lower chance of reaching advanced ages than men in the Indian region of Kerala* or Chinese men. This is despite them having higher incomes even after adjusting for purchasing power. Sen concludes that the difference is due to the wide range of social and economic institutions facing African American men that restrict a basic freedom: "the ability to survive rather than succumb to premature mortality."[7]

Another key aspect of Sen's argument is his suggestion that development organizations—and specifically the World Bank—should adopt a "many-sided approach" to development and not rely on single solutions, such as market liberalization.[8] His approach gives a major role to the state* in ensuring that education, health, and other related freedoms are provided to all in society. The idea is captured in the phrase "human development," which involves the "creation of social opportunities" and the expansion of freedom.[9]

Language and Expression

Sen's ideas in *Development as Freedom* are presented in an accessible way, making the book easy to read for a general audience. Unlike many economics texts, it involves no mathematical formulas. Sen says in the preface that this was indeed his aim: "I have tried to make the discussion as nontechnical as possible, and have referred to the more formal literature—for those inclined in that direction—only in endnotes."[10] In the course of the book, however, Sen does use examples and themes—sometimes without a clear transition—that may at first distract the reader from fully understanding the idea of "human development." Examples of this are his discussion of social choice theory* that is perhaps too abstract for a general audience, and his discussion of the American philosopher John Rawls's* theory of justice.

The fact that the book was based on a series of lectures to employees of the World Bank may explain this lack of focus. As that original audience was deeply involved in the development process, the book sometimes also assumes that the reader too will be familiar with the challenges of development. Nevertheless, general readers will come away from *Development as Freedom* with a clear understanding of Sen's main premise: human-centered development and the expansion of freedom are identical processes.

NOTES

1 Amartya Sen, *Development as Freedom* (New York: Anchor Books, 1999), 3.

2 Sen, *Development as Freedom*, xii.

3 Sen, *Development as Freedom*, 16.

4 Sen, *Development as Freedom*, 17.

5 Sen, *Development as Freedom*, 18.

6 Sen, *Development as Freedom*, 75.

7 Sen, *Development as Freedom*, 24.

8 Sen, *Development as Freedom*, 126.

9 Sen, *Development as Freedom*, 144.

10 Sen, *Development as Freedom*, xiii.

MODULE 6
SECONDARY IDEAS

KEY POINTS

- **Although Sen's secondary ideas are not unique to his work, they help broaden the scope of his argument that freedom is both the means to and the goal of development.**
- **Two of Sen's key secondary ideas are support for democracy as a tool of freedom and his discussion of culture.**
- **Sen's discussion of social choice* (mass elections, for example), and his challenge to the American economist Kenneth Arrow's* impossibility theorem* (according to which no institution other than a dictatorship can combine the opinions of a group into a simple social opinion) have been largely overlooked by readers.**

Other Ideas

Among the many secondary ideas in Amartya Sen's *Development as Freedom*, two are particularly important: the discussion of the benefits and necessity of democracy, and the treatment of culture. These ideas are not unique to Sen; for example, the positive relationship between wealth and democracy that underlies much of his argument was first suggested by the American political sociologist Martin Lipset* in his essay, "Some Social Requisites of Democracy" (1959).[1] It is Sen's incorporation of the idea into an analytical approach founded on the uses and goal of freedom that is original. Similarly, Sen's critique of essentialist* accounts of "Asian values" in his discussion of culture seems to draw heavily on other writers; before *Development as Freedom* was published, the former South Korean president Kim Dae-Jung* had offered a similar critique. Essentialism is the idea that certain

> **“Our conceptualization of economic needs depends crucially on open public debates and discussions, the guaranteeing of which requires insistence on basic political liberty and civil rights.”**
>
> Amartya Sen, *Development as Freedom*

characteristics (cultural attributes, for example) define the features of groups or objects.

Another of the book's secondary ideas is Sen's warning of being too optimistic about the problem of population growth and food security.* Sen notes that food production should be seen as "a result of human agency" and market incentives, not as a natural process.[2] As such, more food could be produced if market conditions made it economically possible. Sen also argues that looking only at food production underestimates the risk of a major catastrophe such as famine, and that policymakers should instead examine the social institutions (both cultural norms and economic institutions such as markets and regulations) that shape society and the production and distribution of food. This is an interesting example of Sen's economic intuitions—in this case, framing the food problem as one of supply and demand—leading to an original perspective on a controversial topic.

Exploring the Ideas

Sen argues in favor of democracy as a political system for three reasons: first, he suggests democracy has "intrinsic importance"; second, it makes "instrumental contributions" to the quality of society; and third, it plays a "constructive role in the creation of values and norms."[3] One of Sen's arguments is that other forms of governance—particularly authoritarian rule (in which government exercises its authority at the expense of the liberty of the individual citizen)—do not seem to be necessary for economic growth and other measures of well-being.

Sen lists a number of "helpful policies" for economic development, including "openness to competition, the use of international markets, a high level of literacy and school education, successful land reforms and public provision of incentives for investment."[4] These all appear to be consistent with democracy. They are also just as possible in nondemocratic societies or regimes. But Sen suggests that the advantages of democracy in terms of personal freedom make it a form of government superior to other available options.

Sen's discussion of culture deals with the presumption that certain cultures may be more favorable to economic development than others. He challenges the idea that so-called "Asian" culture (which he defines broadly to include the traditions following the philosophy of the ancient Chinese philosopher Confucius* and Brahmin* texts, among others) tends toward authoritarianism. As he writes: "The fact is that in any culture, people seem to like to argue with one another, and frequently do exactly that—given the chance."[5] Sen also suggests that Western attitudes to non-Western societies often tend to be "too respectful of authority—the governor, the minister, the military junta, the religious leader."[6] In other words, discussions of these cultural differences are largely driven by an "authoritarian bias" that ignores the way people interact on an everyday basis.[7]

Overlooked

One of the often overlooked arguments made in *Development as Freedom* concerns social choice—the "idea of using reason to identify and promote better and more acceptable societies."[8] Sen's discussion of social choice is heavily influenced by his formal work on the subject, much of it in response to Kenneth Arrow's so-called impossibility theorem, according to which no institution other than a dictatorship can combine the opinions of a group into a simple social opinion. Consider the following voting paradox, often attributed to the pioneering French political scientist Marquis de Condorcet.* There are three options, x, y, z, and three voters, A, B, and C. Suppose person A prefers x to y and y to z, person B prefers y to z and z to x, and person C prefers z to x and x to y. If majority rule is used to pick an option for all of society, then clearly any option with two votes will beat any other option. The paradox is that this vote will lead to "inconsistencies":[9] x beats y by majority rule (through voters A and C), y beats z (through voters A and B) and z beats x (through voters B and C).

The theorem is often interpreted pessimistically: while markets are very good at gathering information, societies are not, because individual preferences are too different.

Sen challenges this interpretation based on the idea of the "informational base."[10] Rather than focusing only on the very fine detail provided in the example above, he suggests looking at a broader range of needs. For example, majority rule only accounts for the *opinions* of the participants, not their *needs*. If the above example also included the information that voter C valued outcome z at $10 million—or, similarly, needed outcome z for some important life function—while voters A and B only valued their most desired outcomes at $5, then the definition of the socially optimal outcome would probably change.

NOTES

1 Seymour Martin Lipset, "Some Social Requisites of Democracy: Economic Development and Political Legitimacy," *American Political Science Review* 53 (1959): 69–105.

2 Amartya Sen, *Development as Freedom* (New York: Anchor Books, 1999), 208.

3 Sen, *Development as Freedom*, 157.

4 Sen, *Development as Freedom*, 150.

5 Sen, *Development as Freedom*, 247.

6 Sen, *Development as Freedom*, 247.

7 Sen, *Development as Freedom*, 247.

8 Sen, *Development as Freedom*, 249.

9 Sen, *Development as Freedom*, 251.

10 Sen, *Development as Freedom*, 279.

MODULE 7
ACHIEVEMENT

KEY POINTS

- **With the publication of *Development as Freedom*, Sen succeeded in changing the debate on development.**
- **Sen benefitted both from the support of the World Bank* and his prestige as a prominent professor in developing this book.**
- **While *Development as Freedom* is presented as a universal work, it has received some isolated criticisms on cultural grounds.**

Assessing the Argument

Sen was extraordinarily successful in achieving his aims for *Development as Freedom*. The book reached a wide audience, and fundamentally altered the debate on economic development. It inspired practical innovations, such as the Human Development Index* (a measure of a society's economic development that incorporates such elements as levels of education, per capita earnings, and average life expectancy) that have been implemented in real-world policy-making. In part, the success of the book is due to Sen's stature as a scholar and his active engagement with the ideas of freedom and capabilities* for several years. Even before the book was published, Sen had attracted a community of scholars dedicated to these subjects.

Despite this success, Sen's work did not go unquestioned. The scholar and journalist Fareed Zakaria,* in a review of the book for the *New York Times*, praises Sen for being "rich in insight and moral intelligence" but then questions his leaps from discipline to discipline: "a little public choice theory here and a little moral philosophy

> **"*Development as Freedom*, which is billed as 'a new general theory of development economics,' is not so much a repudiation of the standard theory as an attempt to synthesize it with a competing one. More than any other of Sen's books, this one is steeped in the ethos of bricolage.* "**
>
> Akash Kapur, "A Third Way for the Third World," *The Atlantic*

there."[1] He goes on: "*Development as Freedom* has neither the comprehensiveness of the best political philosophy nor the elegance of the best economics. It makes one long for a killer theorem."[2] In other words, Sen attempts more than he can accomplish. Still, the ambition of the book is a significant reason for its success, and the prominence of the book in part refutes Zakaria's claim.

Achievement in Context

Sen faced few restrictions in writing and researching *Development as Freedom*. As a prominent academic working in the United States and Europe, he had access to ample funding and protections for academic freedom. And because he wrote the book with support from the president of the World Bank, he was able to reach a wide audience. Indeed, this context may have contributed to the content of the book. In the book's acknowledgements, Sen writes that he "benefited from the opportunity to interact with the staff of the Bank."[3]

Further, since the book is more theoretical than empirical, it was not affected by specific evidence discovered before or after its publication. In fact, Sen bases much of his theory on the work of scholars that predate him by centuries, including the philosopher Aristotle* and the eighteenth-century economic theorist Adam Smith.* That is not to say the book is purely theoretical: Sen's discussion of the differences between African American men and men in the southern Indian state of Kerala, for example, relies on data from both populations.

Today, it is Sen's general work that is most often discussed rather than specifically *Development as Freedom*. His ideas are also repeated in several of his later works, and the debate has moved on to discuss more recent books, such as *Identity and Violence: The Illusion of Destiny* and *The Idea of Justice*.[4]

Limitations

Sen's approach in *Development as Freedom* can be criticized for being diverse and incorporating different viewpoints. Sen cites both Adam Smith and the German economist and political philosopher Karl Marx,* and he writes about markets,* gender issues, and democracy. Different readers are likely to choose the information relevant to their own context and interpret the book in relation to their own work. The book does appear to have struck a universal chord: a book published in Spanish by an interpreter of Sen's thought, *Raíces intelectuales de Amartya Sen: Aristóteles, Adam Smith y Karl Marx* (*Intellectual Roots of Amartya Sen: Aristotle, Adam Smith, and Karl Marx*), for example, comes to the same conclusions about Sen's intellectual influences as those from other cultural backgrounds.[5]

Sen writes from an atheist or secular, nonreligious perspective—but Ashok Singhal* of the World Hindu Council,* an organization created in the 1960s to organize Hindu society around the world, claimed that Sen's work is part of a Christian conspiracy to wipe out Hinduism from India.[6] An isolated opinion, perhaps—but one showing the limits of Sen's universalism, and a reminder that seemingly open-minded ideas such as Sen's can be received very differently by specific audiences.

Development as Freedom has been translated into 11 languages, is available in 54 editions (between 1999 and 2010), and is held by 2,082 libraries worldwide.[7] Much of his other work has not yet, however, been translated from the original English, and so remains partially limited in its potential to cross cultural boundaries.

NOTES

1 Fareed Zakaria, "Beyond Money," *New York Times*, November 28, 1999, accessed September 30, 2015, https://www.nytimes.com/books/99/11/28/reviews/991128.28zakarit.html.

2 Zakaria, "Beyond Money."

3 Amartya Sen, *Development as Freedom* (New York: Anchor Books, 1999), xvi.

4 See Amartya Sen, *Identity and Violence: The Illusion of Destiny* (London: Penguin Books, 2006) and *The Idea of Justice* (Cambridge, MA: Harvard University Press, 2009).

5 Pablo Sánchez Garrido, *Raíces intelectuales de Amartya Sen: Aristóteles, Adam Smith y Karl Marx* (Madrid, Centro de Estudios Políticos y Constitucionales, 2008).

6 R. Sen Sinha and Teresa Nobels, "A Christian Plot: Singhal," *The Indian Express*, December 28, 1998.

7 OCL Center, "WorldCat Identities: Amartya Sen 1933–," accessed October 14, 2015, http://worldcat.org/identities/lccn-n50-12860.

MODULE 8
PLACE IN THE AUTHOR'S WORK

KEY POINTS

- *Development as Freedom* was published in the later stages of Sen's career and builds on all his work.
- Sen's career began with theoretical work on social choice* (a subfield of economics concerned with understanding how the rules of things such as voting, by which preferences are quantified, can translate individual beliefs into rational group action) and evolved to include the philosophical material that makes up this book.
- Sen's principal contribution has been to the field of development economics* (a subfield of economics that describes how nations achieve prosperity by modernizing their economies).

Positioning

Development as Freedom was published when Amartya Sen was well into his sixties and had already won the prestigious Nobel Prize in Economics.* The book can therefore be understood as the final expression of his work on human development. It is not, however, a comprehensive presentation of his academic work, though fragments of many of the subjects that define his career are present in the book; for instance, he dedicates a chapter to women's rights, a chapter to social choice theory, and several chapters to capabilities* (the freedom or capacity of individuals to do the things that make their life meaningful), among other topics. Readers looking for the context for Sen's ideas should seek out his earlier writings after reading this book.

The breadth of Sen's interests is impressive. It is unusual for a scholar, particularly in economics, to migrate from theoretical work to

"The social choice problems that had bothered me earlier on were by now more analyzed and understood, and I did have, I thought, some understanding of the demands of fairness, liberty and equality. To get firmer understanding of all this, it was necessary to pursue further the search for an adequate characterization of individual advantage."

Amartya Sen, Nobel Prize autobiography

broader discussions. Sen takes this process further, building on his theoretical work to create a philosophy of justice, and then adding practical considerations to the discussion. For readers interested in this journey, *Development as Freedom* is undoubtedly the best starting point in Sen's work.

Sen's opinions have changed over time. At the beginning of his professional life (the 1950s and 1960s) he took an anti-market, anti-neoclassical* position—that is, he believed that economic institutions should be regulated by government policy in order to ensure a measure of social justice). From around 1976 onward, he was critical of utilitarianism* (the philosophical position that the best action or governing system is the one that maximizes "utility"*—an abstract measure of well-being). During this period, he brought forward the idea of development as freedom for the individual, which seems to have resolved his conflicted views.

Integration

Sen's career can be seen as having progressed in two stages: in the first, he worked on formal models of social choice; in the second, he drifted into philosophical questions of justice, which eventually led to *Development as Freedom*. In his book *Collective Choice and Social Welfare* (1970), Sen "made an effort to take an overall view of social choice theory."[1] The motivation for this work was a question that had engaged Sen as a child: "Is reasonable social choice at all possible given the differences between one person's preferences and another's?"[2]

Over time, this largely theoretical question gave way to more practical concerns. In a series of lectures in 1979, Sen attempted to develop the concept of "individual advantage."[3] The approach he took "sees individual advantage not merely as opulence or utility, but primarily in terms of the lives people manage to live and the freedom they have to choose the kind of life they have reason to value."[4] It was in the context of this question that Sen first came to the idea of capabilities, which would later appear in *Development as Freedom*.

Following this basic discovery, Sen applied himself to "analyzing the overall implications of this perspective on welfare economics* and political philosophy."[5] Also during this time, he began work on a more practical project—the measurement of human development for the United Nations Development Programme.* Over this period, he published several texts on the causes of famine, among them *Poverty and Famines: An Essay on Entitlement and Deprivation* in 1981[6] and *Hunger and Public Action*, published jointly with Jean Drèze* in 1989.[7] In 1990, he published his article "More than One Million Women Are Missing," highlighting gender bias in Chinese, Indian, Arab, and North African societies.[8] In 1993, he published a book jointly with the American philosopher Martha Nussbaum* entitled *The Quality of Life*,[9] in which he discusses how his capabilities approach is linked to quality of life.

Significance

When Sen was awarded the Nobel Prize in Economics in 1998, the prize committee did not mention capabilities in their explanation for the prize. In fact, Sen received the Prize "for his contributions to welfare economics," which refers to his work on social choice theory.[10] As such, it is difficult to argue that *Development as Freedom* represents the pinnacle of Sen's thought. Still, since many readers will not be familiar with Sen's more formal work on welfare, this book is likely his most influential contribution. Further, while social choice theory has become a relatively inactive area of economics, the capabilities approach described in this book has inspired subsequent work that remains an important part of the development economics conversation. In that way, this book may be seen as more significant than, for example, *Collective Choice and Social Welfare* or Sen's other work.

According to a review of *Development as Freedom* in the British journal the *Economist*: "The distinctive strand in the economic writings of Amartya Sen has been his interest in politics and ethics. In his new book, Mr. Sen … again draws the three disciplines together. *Development as Freedom* is a personal manifesto: a summing up; a blend of vision, close argument, reflection and reminiscence."[11] Sen may or may not agree with this statement—but from the outside looking in it seems appropriate.

NOTES

1 Amartya Sen, "Amartya Sen—Biographical," Nobelprize.org, accessed October 14, 2015, http://www.nobelprize.org/nobel_prizes/economic-sciences/laureates/1998/sen-bio.html. See Amartya Sen, *Collective Choice and Social Welfare* (San Francisco: Holden Day, 1970).

2 Sen, "Amartya Sen—Biographical."

3 Sen, "Amartya Sen—Biographical."

4 Sen, "Amartya Sen—Biographical."

5 Sen, "Amartya Sen—Biographical."

6 Amartya Sen, *Poverty and Famines: An Essay on Entitlement and Deprivation* (Oxford: Oxford University Press, 1981).

7 Jean Drèze and Amartya Sen, *Hunger and Public Action* (Oxford: Clarendon Press, 1991).

8 Amartya Sen, "More Than One Million Women Are Missing," *New York Review of Books*, 37, no. 20 (1990).

9 Amartya Sen and Martha Nussbaum, *The Quality of Life* (Oxford: Oxford University Press, 1993).

10 Nobelprize.org, "The Sveriges Riksbank Prize in Economic Sciences in Memory of Alfred Nobel 1998,", accessed September 30, 2015, http://www.nobelprize.org/nobel_prizes/economic-sciences/laureates/1998/.

11 "The Measure of Progress," *Economist*, September 16, 1999, accessed September 30, 2015, http://www.economist.com/node/239670?zid=295&ah=0bca374e65f2354d553956ea65f756e0

SECTION 3
IMPACT

MODULE 9
THE FIRST RESPONSES

KEY POINTS

- **Sen's critics have challenged him for not making specific policy recommendations, for presenting an overly individualistic view of human capability, and for ignoring group interactions.**
- **Sen has responded directly to his critics, making few concessions but acknowledging that his approach is incomplete.**
- **The debate surrounding *Development as Freedom* has helped to expand the book's reach and scope.**

Criticism

While Amartya Sen's *Development as Freedom* was warmly received on publication, it did meet some criticism; the British scholar Stuart Corbridge,* for example, in a largely positive article discussing "the spaces of Amartya Sen," challenges Sen for not providing specific policy recommendations.[1] Corbridge's critique centers on the idea that Sen's unwillingness to clearly state his policy views limits the overall power of the freedom framework. Corbridge believes that Sen would agree on higher educational standards for all as part of his capabilities approach,* and asks:"But who will object to this nowadays, at least in public, and what, if anything, does it tell us about the policies—and politics—that must be put into place to secure these outcomes?" Corbridge's answer to these questions is "nobody" and "not much."

Another critique comes from the development economics* scholars Frances Stewart* and Séverine Deneulin,* who challenge what they call the "individualism" of Sen's approach to development,

> **"Any book making such grand claims as *Development as Freedom* is bound to engender a measure of debate or even dissent."**
>
> Stuart Corbridge, "*Development as Freedom*: The Spaces of Amartya Sen," *Progress in Development Studies*

saying that, for him, "all social phenomena must be accounted for in terms of what individuals think, choose, and do."[2] This approach, they suggest, ignores the crucial interactions between individuals that make up society.

A more subtle criticism comes from the development scholar Peter Evans* in his essay "Collective Capabilities, Culture, and Amartya Sen's *Development as Freedom*."[3] While mainly congratulatory of Sen, he too challenges Sen's emphasis on the individual, preferring to focus on "collective capabilities." He writes: "Gaining the freedom to do the things that we have reason to value is rarely something we can accomplish as individuals" and argues that collective action is crucial for the expansion of freedom—an idea underdeveloped in Sen's book.[4]

Among other criticisms of Sen's work, the British writer and Africa scholar Alex de Waal* argued that famines are not caused by lack of entitlements to food, but by disease epidemics;[5] feminists have questioned whether Sen adequately dealt with power inequalities between men and women;[6] and environmentalists expressed concern about the lack of an environmental dimension in Sen's works.[7]

General criticism was also voiced by nonacademics. The American newspaper the *Wall Street Journal*, often associated with free-market views, proclaimed that Sen's "renown would outstrip the quality of his work," suggesting that the author's fame was due not to the substance of his work but rather to his embrace by political liberals (who, unlike economic liberals, are generally opposed to the unregulated markets).[8]

Responses

Sen responded directly to some of his critics (particularly those who said his view was too individualistic) in the essay "Response to Commentaries," published in 2002.[9] Sen rejects Stewart and Deneulin's charge of individualism, arguing instead that his theoretical approach is fully compatible with a social one. He writes: "No individual can think, choose, or act without being influenced in one way or another by the nature and working of the society around him or her."[10] Sen argues that the approach he described in *Development as Freedom*, particularly his discussion of the benefits of democracy, fully includes group dynamics. As he says: "To note the role of 'thinking, choosing, and doing' by individuals is just the beginning of a manifest reality, but we cannot end there without an appreciation of the deep and pervasive influence of society on such 'thinking, choosing, and doing.'"[11]

In response to Peter Evans, Sen rejects the notion of "collective capabilities." He argues that even if interactions with others play a crucial role in human life, the capabilities that bring about those actions are "*socially dependent individual capabilities*."[12] The implication is that, while group characteristics matter, the individual is still the thing worth examining. The "intrinsic satisfactions that *occur in* a life must occur in an individual's life, but in terms of causal connections, they *depend* on interactions with others."[13] Sen distinguishes between what he sees as individual capabilities and actual collective capabilities—that is, the freedom or capacity of groups to respond collectively to threats. An example of a collective capability would be a country's arsenal in response to nuclear attack.

Conflict and Consensus

These conversations inspired by *Development as Freedom* and Sen's earlier work have continued beyond the first exchanges described above. For example, in a 2005 paper titled "Groups and Capabilities"

Frances Stewart argues for greater inclusion of group concepts into the capabilities approach pioneered in *Development as Freedom*.[14] Exactly what is the right balance of groups versus individuals in working out capabilities has not been resolved. In fact it is probably only of interest to scholars trying to develop the capabilities approach into a rigorous philosophical discipline. From a practical point of view, these differences are unlikely to matter.

There has been a lively debate about Sen's freedom perspective on development since the book's publication, which has stimulated discussions on poverty, development, and welfare. *Development as Freedom* is a stepping-stone for a new framework for development based on capabilities and human development. As such, much of the exchange between Sen and his critics has taken place *within* the human development perspective that Sen pioneered. This may relate to Corbridge's criticism that if Sen's ideas are difficult to disagree with, they are also difficult to implement. It is accepted among development economists that an understanding of human capabilities and freedom is essential for doing development properly; the details of the approach, however, remain a matter of debate.

NOTES

1 Stuart Corbridge, "*Development as Freedom*: The Spaces of Amartya Sen," *Progress in Development Studies* 2, no. 3 (2002): 183–217.

2 Frances Stewart and Séverine Deneulin, "Amartya Sen's Contribution to Development Thinking," *Studies in Comparative International Development* 37, no. 2 (2002): 66.

3 Peter Evans, "Collective Capabilities, Culture, and Amartya Sen's *Development as Freedom*," *Studies in Comparative International Development* 37, no. 2 (2002): 54–60.

4 Evans, "Collective Capabilities," 56.

5 Alex de Waal, "Famine Mortality: A Case Study of Darfur, Sudan 1984–85," *Population Studies* 43, no. 1 (1989): 5–24.

6 Marianne Hill, "Development as Empowerment," *Feminist Economics* 9 (2003): 117–35.

7 Amartya Sen et al., "Continuing the Conversation," *Feminist Economics* 9 (2003): 319–32.

8 Robert L. Pollock, "The Wrong Economist Won," *Wall Street Journal*, October 15, 1998, accessed November 10, 2015, http://www.wsj.com/articles/SB908405798502442000.

9 Amartya Sen, "Response to Commentaries," *Studies in Comparative International Development* 37, no. 2 (2002): 78–86.

10 Sen, *Response*, 81.

11 Sen, *Response*, 81.

12 Sen, *Response*, 85.

13 Sen, *Response*, 85.

14 Frances Stewart, "Groups and Capabilities," *Journal of Human Development* 6, no. 2 (2005): 185–204.

MODULE 10
THE EVOLVING DEBATE

KEY POINTS

- **Sen's capabilities* approach has been applied to several problems in the field of development economics,* and has been used to generate measures of development that include a broad set of factors.**
- ***Development as Freedom* is a key text in the field of human development, an approach to development based on capabilities.**
- **One of the key scholars to develop Sen's ideas is the American philosopher Martha Nussbaum,* who has attempted to formalize the capabilities framework in philosophical terms.**

Uses and Problems

The ideas contained in *Development as Freedom* have had a deep influence on work in development economics and policy-making, as confirmed by the Human Development Reports* created by the United Nations Development Programme*—a branch of the United Nations dedicated to global development, providing both direct assistance and advice to poor countries. Though these reports predate *Development as Freedom*, they have over time included elements of Sen's capabilities approach such as human rights* (2000), democracy (2002), and climate change* (2007–8), among others. The reports have also evolved in terms of their measurements to include an inequality-adjusted Human Development Index, or HDI* (a measure of a society's economic development that incorporates such elements as levels of education, per capita earnings, and average life expectancy), a Multidimensional Poverty Index, or MPI* (an index based on factors such as child mortality, years of schooling, and various aspects of living

> **“Sen’s arguments stem from a commitment to the importance of individual freedoms. Not for him the wishy-washy relativism that gripped many parts of the academy in the 1990s. But not for him either the bone hard individualism of the intellectual right. This means we need to be careful in interpreting Sen.”**
>
> Simon Reid-Henry, “Amartya Sen: Economist, Philosopher, Human Development Doyen,” *Guardian*

standards to measure poverty), and a Gender Inequality Index, or GII* (an index used to measure gender disparity, including variables on reproductive health, labor market participation, and female empowerment).

The HDI is particularly important. It ranks countries based on their development performance, and is often cited by policymakers when discussing potential directions for policy. As a matter of comparison, in the 2014 HDI, Norway was ranked first, the United States fifth, and Niger last.[1] These measures are used by governments and international organizations to make decisions about how resources are allocated. The inclusion of capabilities measures clearly shows the influence of Sen’s work.

The Human Development Reports have also influenced the bringing forward of the Laeken indicators,* a set of measures of poverty and social exclusion used by the European Union.* These include most of the traditional measures of development, such as income and inequality, but also include education and health measures that seem to come from Sen’s capabilities approach.[2]

Further, the capabilities approach has served as a theoretical framework for official poverty and wealth reports created by the German government. In 2005, Germany decided to combine the previous conditions of life approach* (*Lebenslagen-Ansatz*—an approach to measuring and characterizing poverty, introduced in Germany in 1917) with Sen's capabilities approach.[3] Germany was thus the first country to introduce poverty reports based on capabilities. Other countries have begun to follow, including Great Britain, the United States, Italy, and the Netherlands.

Schools of Thought

Development as Freedom has enjoyed a wide audience, and scholars from many disciplines, such as economics, philosophy, gender studies, development studies, human rights, and health science, have been inspired by Sen's ideas. If there is one single school of thought that best exemplifies the text, however, it would be the human development school. The closely related concepts of human development and capabilities are often assumed to be synonymous—judging by *Development as Freedom*, Sen considers them so. Other scholars also tend to combine the processes of expanding freedom, building capabilities, and achieving human development.

Rather than being based just on *Development as Freedom*, the human development school emerged from Sen's work as a whole. The school is not formally established—there is no physical school that exclusively teaches human development models and approaches. But the existence of an association centering on human development, a journal, wide press coverage on the topic, honorary lectures in Sen's name, and support from academic followers across disciplines, all indicate the existence of a "school." Perhaps the best evidence of this is the Human Development and Capability Association (HDCA),* founded in 2004 to promote "high quality research in the interconnected areas of human development and capability."[4] The first president of the HDCA was Sen himself, and subsequent presidents have been his close collaborators, including Martha Nussbaum and the development economics professor Frances Stewart.* Another important outlet for ideas linking development and capabilities is the *Journal of Human*

Development and Capabilities, which publishes academic work spanning the many disciplines touched by the field.

In Current Scholarship

Perhaps the most notable scholar to develop Sen's ideas beyond *Development as Freedom* is Martha Nussbaum, who has collaborated with Sen throughout his career. Nussbaum, who trained as a philosopher at Harvard, writes in a more orthodox philosophical style than Sen, focusing on theory and completeness. She distances her and Sen's work from the application of those ideas by organizations such as the United Nations, writing: "[the UN reports on human development] use the notion of capabilities as a comparative measure rather than as a basis for normative political theory. Amartya Sen had a major intellectual role in framing them, but they do not incorporate all aspects of his (pragmatic and result-oriented) theory."[5] As Sen and Nussbaum have often worked together, their ideas are frequently considered together; readers should therefore be careful to separate the contributions of Nussbaum from Sen's original thought.

In order to fill some of the intellectual gaps in the capabilities approach, Nussbaum has developed a systematic and influential capability theory rooted in the ideas of Aristotle.* She has also developed a list of fundamental capabilities and a threshold for achieving those capabilities. The list includes life, bodily health, bodily integrity, senses, imagination and thought, emotions, practical reason, affiliation (being able to live with other people), other species (being able to live in the company of other animals), play, and control over one's environment.[6]

Frances Stewart, who was critical of aspects of *Development as Freedom*, has challenged Nussbaum, suggesting her list does not form a contemporary "overlapping consensus" of the concepts included in a capabilities framework.[7] Stewart, along with the economist Paul Streeten,* was involved in the development of the basic needs approach,* an alternative model to the capabilities approach for measuring and understanding poverty focused on defining the minimum resources needed for living a comfortable life—typically defined in terms of consumption goods.[8]

NOTES

1 United Nations Development Programme, "2014 Human Development Index," accessed September 30, 2015, http://hdr.undp.org/en/content/table-1-human-development-index-and-its-components.

2 Poverty.org, "Laeken Indicators," accessed September 30, 2015, http://www.poverty.org.uk/summary/eu.htm.

3 Christian Arndt and Jürgen Volkert, "The Capability Approach: A Framework for Official German Poverty and Wealth Reports," *Journal of Human Development and Capabilities* 12, no. 3 (2011): 311–37.

4 Human Development and Capability Association, "HDCA Mission and History," accessed September 30, 2015, https://hd-ca.org/about/hdca-history-and-mission.

5 Martha Nussbaum, *Creating Capabilities: The Human Development Approach* (Cambridge, MA: Harvard University Press, 2001), 17.

6 Nussbaum, *Creating Capabilities*, 30–1.

7 Frances Stewart, "Women and Human Development: *The Capabilities Approach*, by Martha Nussbaum," *Journal of International Development* 13, no. 8: 1191–2.

8 Paul Streeten, Shahid Javed Burki, Ul Haq, Norman Hicks, and Frances Stewart, *First Things First: Meeting Basic Human Needs in the Developing Countries* (New York: Oxford University Press, 1981).

MODULE 11
IMPACT AND INFLUENCE TODAY

KEY POINTS

- *Development as Freedom* still influences policy-making and remains a core text in many courses on development.*
- The intellectual debate on human development has drifted into many areas, including historical and macroeconomic trends and measurement; "macroeconomic trends" refers to trends in large-scale economic features such as international trade and national employment rates.
- The human development school has been contrasted with the human rights school* (the position that human rights* are a central component in the expansion of human prosperity). Sen's capabilities* argument has been used to strengthen the rights school.

Position

Although thinking around development has produced alternatives to the capabilities approach—including exploration of topics such as demography (human population statistics), industrial policy, and the environment—Amartya Sen's ideas, as they are expressed in *Development as Freedom*, remain influential. Perhaps the most notable example of this continuing influence is the application of capabilities ideas to real-world policy-making, as represented by measures such as the Human Development Index.*

Development as Freedom is also a key text in many undergraduate courses on development economics.* For example, a syllabus for a course on "Development Economic Policy" at the Harvard Kennedy School of Government calls it a "classic" book on international development, alongside other pathbreaking books on the subject. But

> **"To promote the good life, a capability and health account values longevity and freedom from disease. It emphasizes prevention and treatment, favoring those most deprived in health and at risk of health deprivation. It also emphasizes individual agency and supports efforts to improve health to equip individuals with the mental and physical ability required for agency."**
>
> Jennifer Prah Ruger, *Toward a Theory of a Right to Health: Capability and Incompletely Theorized Agreements*

while students still read the book, Sen's work is not at the cutting edge of the development field; although the capabilities approach helped to add a human dimension to development, later work has pushed the field well beyond Sen's premise.

Specifically, development economics has become a much more empirical discipline than *Development as Freedom* might suggest (that is, it has moved from the realm of theory to the realm of the measurable and material). This empirical shift can, in fact, be traced back to Sen; by expanding the scope of development analysis, he opened a door to new problems to be identified and measured. Much literature is now dedicated to measuring education quality in developing countries, for example; the economist Esther Duflo* has been a leader in this effort, using randomized controlled experiments (a means to obtain secure results by comparing cases where certain criteria either do or do not apply) to identify the benefit of particular interventions.

Interaction

As it continues to expand in scope and splinter into differing views, the human development debate remains active. The *Journal of Human Development and Capabilities*, originally founded to offer a platform for Sen's approach, still draws prominent thinkers to write about the topic (among them the Nobel Prize-winning economist Joseph Stiglitz* in 2012). A recent edition of the journal focused on predictions of human development in Latin America over the next one hundred to two hundred years. Further, the search to connect human development with other topics continues, as research on human development and macroeconomics, measurement, children, and education shows.

Despite moving away from the intellectual cutting edge, *Development as Freedom* is still used as a starting point for many research initiatives. One important and unsolved area of work concerns how best to *measure* human development—indeed, the wide use of development measures by international organizations has made this more than an academic matter.[1] Interestingly, the discussion around measurement is in some ways a departure from the spirit of *Development as Freedom*. Sen himself did not believe that a crude index of, for example, three components of life could fully capture the complexity of the process of human development. As he later revealed in an interview, Sen reluctantly agreed to assist with developing the Human Development Index with the economist Mahbub ul Haq,* despite seeing it as a "vulgar" measure.[2] Sen almost certainly used the term "vulgar" ironically, and was simply suggesting that translating a full model of human development into numbers is full of challenges.

The Continuing Debate

One continuing debate about the human development perspective concerns the human rights school and the role of rights versus capabilities. The conflict between the two frameworks arises when broad rights describe a desired state of the world but not the actual steps needed to achieve it. For example, the scholar Jennifer Prah Ruger* has been a strong advocate for a universal right to health but has also worked to develop *a theory* of the right to health. Ruger notes the challenge of creating a unified approach to rights: "While activists, non-governmental organizations, and scholars have made significant progress in promoting a human rights approach to health and the field of health and human rights more generally, the question of a philosophical and conceptual foundation—a theory—for the right to health has fallen through the cracks."[3]

One of the problems in developing a theory of the human right to health is the breadth of disciplines and questions related to the meaning and reality of health. Health involves not just the legal frameworks for providing health care, and the means of providing health in society, but also broader considerations of the general health of a population. In order to unify these perspectives into a single theory of the right to health, Ruger draws heavily on Sen's capabilities approach. As she writes: "To operationalize a right to health [that is, roughly, to determine what this abstract right signifies in a material, measurable sense], determining how to measure capabilities and at what level to provide them is necessary."[4] The interplay between rights and capabilities is important: in order to develop a broad theory of the right to health, Ruger used Sen's ideas as a tool to clarify the parameters of such a right.

NOTES

1 Georges Nguefack-Tsague, Stephan Klasen, and Walter Zucchini, "On Weighting the Components of the Human Development Index: A Statistical Justification," *Journal of Human development and Capabilities* 12, no. 2 (2011): 183–202.

2 David Hamburg, "Preventing Genocide: Transcript of Interview with Amartya Sen," accessed October 20, 2015, https://lib.stanford.edu/preventing-genocide/transcript-interview-amartya-sen.

3 Jennifer Prah Ruger, "Toward a Theory of a Right to Health: Capability and Incompletely Theorized Agreements," *Yale Journal of the Law and Humanities* 18, no. 2 (2006): 273.

4 Ruger, "Right to Health," 303.

MODULE 12
WHERE NEXT?

KEY POINTS

- Despite some radical critiques that challenge the idea of development* altogether, Development as Freedom will be remembered as a classic in the field.
- One example of the future direction of development research is the work of the Abdul Latif Jameel Poverty Action Lab,* an institution at the Massachusetts Institute of Technology dedicated to using scientific methods to understand the best means of alleviating poverty, which incorporates human development components into rigorous methodologies.
- *Development as Freedom* is a seminal book for those wanting an entry point into the debates that make up the field of development economics.*

Potential

Amartya Sen's *Development as Freedom* is likely to be remembered as a classic in the field of development economics, and the direction of the discipline suggests future work will be built in some way on Sen's ideas. It is very unlikely that a development scholar today would dispute the importance of human capabilities*—though some might challenge the relative importance of various measures of capability. One exception to this is post-development theory,* which rejects the process of development and argues that it was "invented" by wealthy countries to suppress the threat of socialism in poor countries.[1] This is a radical critique, however, and is unlikely to derail the overall development economics project of which Sen's work is a part.

> **“Freedom has a thousand charms to show,
> That slaves, howe'er contented, never know.”**
> William Cowper, quoted in *Development as Freedom*

While it is hard to predict the specific ways in which Sen's ideas will be further developed, some guesses can be made. First, it is likely that the list of capabilities considered essential for life will expand and capabilities relating to an individual's right to self-identity may be added to the lexicon. Second, development economics will continue to become more empirical, and measurement will become more exact. Current measures such as the Human Development Index* translate the many factors that constitute human life into a single number for each country. But technological innovations and new data collection techniques will allow the breadth of these measures to expand dramatically. For example, it is reasonable to expect that so-called biometric* tools (sensors that measure an individual's physiological performance) will make measurements of health capabilities much clearer.

Future Directions

Development economics is a large and growing field, and many of its aspects draw in some way on Sen's work. In fact, development today is by definition human development. One interesting initiative in the field is the creation of the Abdul Latif Jameel Poverty Action Lab (J-PAL)* at the Massachusetts Institute of Technology (MIT), led by the French economist Esther Duflo.* The cornerstone of the J-PAL approach is to implement rigorous methods, such as randomized controlled experiments,* to determine which policy interventions work, and by how much they are more successful than alternatives.

One notable scholar affiliated with J-PAL is David Atkin,* also a professor at MIT. Atkin's research profile states that he studies the "impacts of trade liberalization on the poor in the developing world" and "the role of regional taste differences in altering the impacts of trade reforms in India, and educational responses to the rise of export oriented manufacturing in Mexico."[2] One thing that stands out from this description is its breadth—it is hard to imagine a development scholar 50 years ago focusing on several regions separated by oceans. Today, with more rigorous data collection techniques worldwide, working on multiple problems around the globe is possible. Atkin's profile also reveals the extent to which development research today is human development; he deals with traditional subjects such as trade, but also education and health.

Summary

Development as Freedom helped to develop a perspective on development that pushed beyond traditional utilitarian* models—models focused on outcomes—and expanded the scope of economic analysis. As Sen himself says: "I have tried to argue for some time now that for many evaluative purposes, the appropriate 'space' is neither that of utilities … nor that of primary goods (as demanded by Rawls),* but that of the substantive freedoms—the capabilities—to choose a life one has reason to value."[3]

This is important for both practical and conceptual reasons. Conceptually, it heightens the moral power of economic analysis, and provides a systematic framework for comparing different ways to allocate goods and rights in society. And practically, it challenges policy-makers, scholars, and communities to direct their action toward the expansion of freedom.

In this way, *Development as Freedom* is both textbook and manifesto. Readers of the book will be exposed to an influential perspective on development. But they will also be challenged to think deeply about some of the most important aspects of human existence. Those with an interest in economics will be challenged to consider how economic ideas and models can be compatible with a rich understanding of human welfare. By pushing readers away from reductionist explanations of development and welfare—that is, from explanations of development and welfare intentionally stripped of the complexity of the "real world"—the book will open and challenge minds from a variety of disciplines.

NOTES

1 InterAmerican Wiki, "Post-Development," accessed September 30, 2015, http://wiki.elearning.uni-bielefeld.de/wikifarm/fields/ges_cias/field.php/Main/Unterkapitel163.

2 David Atkin, "David Atkin," accessed September 30, 2015, http://www.povertyactionlab.org/atkin.

3 Amartya Sen, *Development as Freedom* (New York: Anchor Books, 1999), 74.

GLOSSARY

GLOSSARY OF TERMS

Abdul Latif Jameel Poverty Action Lab (J-PAL): an institution at the Massachusetts Institute of Technology dedicated to using scientific methods to understand the best methods for alleviating poverty.

Basic needs approach: an approach to measuring poverty focused on defining the minimum resources needed for living a comfortable life. These needs are typically defined in terms of consumption goods.

Bengal famine of 1943: a catastrophe that led to the deaths of around three million people and significant social disruption in India. Many of the deaths associated with the famine were due to disease rather than a lack of food.

Brahmin: member of a community of people in India, sometimes considered the highest of the major castes of Indian culture. Historically, brahmins were scholars and spiritual leaders.

Bricolage: something made from various available materials. "Bricolage" is the French for do-it-yourself home improvement projects. In his article about Sen in *The Atlantic*, Akash Kapur linked the term to French anthropologist Claude Lévi-Strauss (1908–2009) and defined it as "the drawing together of diverse traditions and ideas into a new reality."

Capabilities: a concept used throughout *Development as Freedom* describing the capacity of individuals to do the things that make their life meaningful. An example of a capability is access to shelter. A capability set is the maximum amount of capability available to a person.

Capitalism: an economic system in which transactions are carried out through markets, facilitated by the investment of capital by private individuals.

Climate change: the observation that an increase of greenhouse gases in the Earth's atmosphere due to human activity is increasing atmospheric temperatures with unknown future consequences.

Cold War: a global "tension" lasting from 1947 to 1991 between Western bloc countries and the communist countries of the Eastern bloc, most notably the Soviet Union. The tension occasionally led to proxy wars in places such as Vietnam.

Conditions of life approach: an approach to measuring and characterizing poverty introduced in Germany in 1917 by Otto Neurath. It bears striking similarities to Sen's capabilities approach and is opposed to methods that do not allow for interpersonal comparisons of utility (as is the case in utilitarianism).

Currency devaluation: the process of reducing the value of a country's currency relative to the currencies of the country's trading partners. In theory, devaluation tends to make exports more affordable and can increase economic growth.

Development: in the study of economics, the process by which countries transition from premodern economic systems to dynamic economies.

Development economics: a subfield of economics that describes how countries transition from premodern to modern economies and how chronic conditions of poverty can be alleviated.

Essentialism: the idea that certain characteristics (certain cultural attributes, for example) define the features of groups or objects. An example is the idea held by some that Asian cultures are not conducive to democracy.

European Union: the political and economic union of 28 member states established by the Maastricht Treaty of 1993. One important feature of the union is the creation of a single market among members.

Fabian Society: a British social organization, founded in 1884, dedicated to promoting greater equality and opportunity. The society is considered to promote a socialist economic perspective.

Food security: the ability to access enough safe, nourishing, and affordable food for a group of people to live a healthy, active life.

Functionings: the abilities and behaviors that make up a life. For example, eating is a functioning—as is spiritual fulfillment.

Gender Inequality Index (GII): an index used to measure gender disparity as part of the Human Development Report. It includes variables on reproductive health, labor market participation, and female empowerment, among other factors.

Harvard Society of Fellows: a networking organization designed to develop young scholars in a variety of disciplines.

Human Development and Capabilities Association (HDCA): a global organization consisting of academics and policy-makers dedicated to understanding poverty, justice, welfare, and economics.

Human Development Index (HDI): a measure of a society's economic development that incorporates such elements as levels of education, per capita earnings, and average life expectancy.

Human Development Report: an annual publication of the United Nations Development Programme that measures various aspects of economic development, including income, environmental sustainability, gender equality, and education—among other factors.

Human rights: a set of moral principles describing standards of human behavior that are protected by international law. The concept is controversial because some disagree whether or not such things as education are rights.

Human Rights School: a philosophical perspective that considers the recognition of human rights by national and international governing bodies as a central component of expanding human prosperity.

Impossibility theorem: a result in social choice theory first proved rigorously by the economist Kenneth Arrow showing that all systems for aggregating preferences (majority rule, for example) other than dictatorship may lead to inconsistent outcomes, sometimes known as cycles.

International Monetary Fund (IMF): an international organization founded in 1945 with the goal of expanding international trade and the exchange of goods and services between countries. The IMF is often a lender of last resort in global financial crises.

Kerala: a state in southern India notable for its unique combination of low income per capita and a high degree of human development.

Keynesian economics: a collection of ideas associated with the British economist John Maynard Keynes. Keynesians traditionally believe that the government can help to stabilize an economy in the event of financial crises by supplementing private spending with public spending.

Laeken indicators: a set of indicators on poverty and social exclusion adopted by the European Union. They include such measurements as the poverty rate, regional cohesion, life expectancy, and educational factors, among others.

Libertarianism: an ideology based on the idea that individual liberty is central to all political and economic life; libertarians tend to advocate for less government intervention in the economy.

Liberalization: the process of removing restrictions, most often imposed by governments, on market exchange. An example of liberalization would be the removal of tariffs on trade between countries.

Markets: the institutions that organize economic behavior between willing sellers of goods and willing buyers. An alternative economic system would be to allow governments to set prices.

Market exchange: the interactions between buyers and sellers in a market. A fundamental notion of exchange is that in order for an exchange to occur, both the buyer and seller must be at least as well off after the exchange as before.

Multidimensional Poverty Index (MPI): an index developed in 2010 by the United Nations Development Programme and the Oxford Poverty and Human Development Initiative that uses factors such as child mortality, years of schooling, and various aspects of living standards to measure poverty.

Neoclassical economics: an approach to economics, often contrasted with Keynesian economics, that seeks to understand economic behavior through an analysis of market institutions and the establishment of prices.

Neoliberalism: a term often used negatively to describe a perspective on economic development that emphasizes the liberalization of markets and structural reforms associated with the Washington Consensus.

Nobel Peace Prize: a prize awarded annually by the Norwegian Nobel Committee for work that fosters global cooperation and peace. Created in 1901, the prize has historically been one of the most controversial Nobel prizes because it is often awarded to political figures, including Barack Obama and Henry Kissinger.

Nobel Prize in Economics: also known as the Nobel Memorial Prize in Economic Sciences, this is an annual prize awarded for work that advances theoretical or empirical understanding in economics. It is considered to be one of the highest honors in the discipline.

Opportunity aspect of freedom: refers to the ability to actually achieve freedoms in society. For example, if the freedom we are interested in is economic freedom, the opportunity aspect would refer to whether people can actually participate in economic life by being able to act as consumers, enter into employment contracts, and start businesses.

Post-development theory: a collection of ideas from scholars who reject the concept of economic development altogether as a Western invention.

Process aspect of freedom: refers to the ability to perform certain aspects of freedom within the rules of a society. For example, the right to vote affirms a particular process freedom.

Randomized controlled experiments: an experimental approach to social science research in which one group is randomly assigned some treatment, and another group is assigned a control. This is considered one of the best ways to identify scientifically the effects of policy interventions on individual outcomes.

Social choice theory: a subfield of economics concerned with understanding how preference aggregation rules, such as voting, can translate individual beliefs into rational group action. Kenneth Arrow's impossibility theorem is a key result.

State: refers to the organization of government into a single entity; states vary in the extent to which they intervene in economic affairs, from complete control to *laissez-faire* (a very minimal degree of government intervention).

Unfreedoms: a term coined by Sen to designate things, either man-made or natural, that prevent us from living the life we want to live: for example, lack of access to health care, sanitation, or basic political and civil rights or lack of nutrition due to famine.

United Nations Development Programme: a branch of the United Nations dedicated to global development. The UNDP provides both direct assistance and advice to poor countries.

Utilitarianism: a philosophy based on the idea that the best action or governing system is the one that maximizes utility, where utility is an abstract measure of well-being.

Utility: an abstract measure of well-being used by economists to compare the outcomes of policies or institutional designs.

Washington Consensus: a perspective on economic development revolving around 10 specific policies promoted by international organizations such as the IMF and World Bank. The policies include reforms, such as currency devaluation and spending cuts, designed to make markets function more efficiently.

Wealth: the measure of the resources and value in a society's economy derived from both material and intellectual sources.

Welfare economics: a subfield of economics concerned with how economic well-being is created and distributed throughout society.

World Bank: an international organization founded in 1944 with the explicit goal of reducing global poverty. The World Bank has played a central role in implementing economic reforms in developing countries since its creation.

World Hindu Council: an organization created in the 1960s to consolidate Hindu society through education and health programs.

PEOPLE MENTIONED IN THE TEXT

Aristotle (384–322 B.C.E.) was a Greek philosopher. A student of the philosopher Plato, he is one of the most famous philosophers in history and has influenced the development of the academic disciplines of ethics and logic, among many others.

Kenneth Arrow (b. 1921) is an American economist, winner of the Nobel Prize in Economics together with John Hicks in 1972. He has contributed to social choice theory and general equilibrium analysis, among other mathematical topics.

David Atkin is an economist currently teaching at the Massachusetts Institute of Technology. His work spans a number of topics in development economics, from health and education to trade.

Saint Augustine (354–430) was a Christian philosopher whose works *The City of God* and *Confessions* are considered major contributions to Western philosophy.

Marquis de Condorcet (1743–94) was a French philosopher and mathematician who was one of the first to conceive of a formal approach to political science.

Confucius (551–479 B.C.E.) was a Chinese philosopher known as the intellectual father of the philosophy of Confucianism. Confucianism is both a philosophy and a set of social and political teachings.

Stuart Corbridge (b. 1957) is a British academic specializing in development studies and geography based at the London School of Economics.

Kim Dae-Jung (1925–2009) was the eighth president of South Korea, serving from 1998 to 2003.

Gérard Debreu (1921–2004) was a French economist who was instrumental in applying mathematical methods to economics. He is considered the first to have developed a theory of general equilibrium in economics.

Séverine Deneulin (b. 1974) is a lecturer in development economics at the University of Bath, and a fellow of the Human Development as Capability Association.

Jean Drèze (b. 1959) is a development economist who has written on famine, gender inequality, health care, and other topics, with a specific focus on India.

Esther Duflo (b. 1972) is a French economist and winner of the prestigious John Bates Clark Medal who mainly works in the field of development economics. She is the cofounder of the Abdul Latif Poverty Action Lab.

Peter Evans is senior fellow in international and public affairs at the Watson Institute, Brown University in Providence , Rhode Island, as well as professor *emeritus* in sociology at the University of California, Berkeley.

Mahbub ul Haq (1934–98) was a Pakistani economist who assisted with the creation of the human development theory and the Human Development Index.

John Maynard Keynes (1883–1946) was a British economist who had a major influence on academic economics and policy. He is perhaps best known for describing how governmental interventions in the economy during times of crisis may reduce the magnitude of the crisis.

Martin Lipset (1922–2006) was an American sociologist who wrote about democratic transitions and was a fellow at the Hoover Institution at Stanford University.

Karl Marx (1818–83) was a German philosopher, economist, historian, sociologist, and revolutionary socialist. He published many books, most importantly *The Communist Manifesto* (1848) and the three volumes of *Capital* (1867, 1885, 1894).

Robert Nozick (1938–2002) was an American philosopher who is known for providing the intellectual foundations of libertarian thought.

Martha Nussbaum (b. 1947) is an American philosopher who has contributed substantially to the capabilities approach to human development, as well as feminism and animal rights. She currently holds a joint appointment in philosophy and law at the University of Chicago.

Plato (424–348 B.C.E.) was a major Greek philosopher and student of Socrates who made contributions to a number of areas, including political philosophy and ethics. His most famous student was Aristotle.

John Rawls (1921–2002) was a leading moral and political philosopher. His most influential work was *A Theory of Justice*, published in 1971. Later, he extended his work by introducing "the idea of an overlapping consensus," which he explained in an article of that title published in the *Oxford Journal of Legal Studies in 1987*.

Jennifer Prah Ruger is a health policy scholar based at the University of Pennsylvania. Ruger works on health equity and the provision of health care in poor countries, and she has developed a theory of the right to health that draws upon Sen's capabilities approach.

Ashok Singhal (b. 1926) was the president of the World Hindu Council for more than 20 years, until 2011.

Adam Smith (1723–90) was a moral philosopher and one of the first modern economic theorists, born in Scotland. His best-known book is *An Inquiry into the Nature and Causes of the Wealth of Nations* (1776). It emphasizes that individual liberty and self-interest are beneficial for society.

T. N. Srinivasan (b. 1933) is an Indian economist whose work covers the topics of development economics, trade, and statistics. He served as adviser to the World Bank from 1977 to 1980.

Frances Stewart (b. 1940) is professor of development economics at the University of Oxford. She was the president of the Human Development and Capability Association in 2008–10.

Joseph Stiglitz (b. 1943) is an American economist, professor at Columbia University. He won the Nobel Prize in Economics in 2001. He is a former chief economist to the World Bank and chairman of the US president's Council of Economic Advisers.

Paul Streeten (b. 1917) is an economist who was a professor at Boston University. He was also director of the Institute of Development Studies (IDS) at the University of Sussex, deputy director-general at the Ministry of Overseas Development, and special adviser to the World Bank. He was the founder of the basic needs approach before he started using Sen's human development approach in his articles.

Alex de Waal (b. 1963) is a British social anthropologist and executive director of the World Peace Foundation at the Fletcher School of Law and Diplomacy at Tufts University. He has written about such diverse topics as famine, humanitarianism, peace-building, human rights, and HIV/AIDS.

John Williamson (b. 1937) is a British economist who coined the phrase Washington Consensus. Williamson spent much of his academic career at the Peterson Institute for International Economics, a macroeconomic policy think tank.

James Wolfensohn (b. 1933) was the ninth president of the World Bank, serving from 1995 to 2005.

Fareed Zakaria (b. 1964) is a journalist and writer who currently hosts a television news show on CNN. He holds a PhD in politics from Harvard University.

WORKS CITED

WORKS CITED

Aristotle. *The Nicomachean Ethics*. Translated by David Ross and revised by J. O. Urmson. Oxford: Oxford University Press, 1980.

Arndt, Christian, and Jürgen Volkert. "The Capability Approach: A Framework for Official German Poverty and Wealth Reports." *Journal of Human Development and Capabilities* 12, no. 3 (2011): 311–37.

Atkin, David. "David Atkin." Accessed September 30, 2015. http://www.povertyactionlab.org/atkin.

Corbridge, Stuart. "*Development as Freedom*: The Spaces of Amartya Sen." *Progress in Development Studies* 2, no. 3 (2002): 183–217.

Drèze, Jean, and Amartya Sen. *Hunger and Public Action*. Oxford: Clarendon Press, 1991.

Evans, Peter. "Collective Capabilities, Culture, and Amartya Sen's *Development as Freedom*." *Studies in Comparative International Development* 37, no. 2 (2002): 54–60.

Garrido, Pablo Sánchez. *Raíces intelectuales de Amartya Sen: Aristóteles, Adam Smith y Karl Marx*. Madrid: Centro de Estudios Políticos y Constitucionales, 2008.

Hamburg, David. "Preventing Genocide: Transcript of Interview with Amartya Sen." Accessed October 20, 2015. https://lib.stanford.edu/preventing-genocide/transcript-interview-amartya-sen.

Hill, Marianne. "Development as Empowerment." *Feminist Economics* 9 (2003): 117–35.

Human Development and Capability Association. "HDCA Mission and History." Accessed September 30, 2015. https://hd-ca.org/about/hdca-history-and-mission.

Huxley, T. H. *Science and Culture and Other Essays*. London: Macmillan and Co., 1888.

InterAmerican Wiki. "Post-Development." Accessed September 30, 2015. http://wiki.elearning.uni-bielefeld.de/wikifarm/fields/ges_cias/field.php/Main/Unterkapitel163.

Kapur, Akash. "A Third Way for the Third World." *The Atlantic*, December 1999. Accessed November 11, 2015. http://www.theatlantic.com/magazine/archive/1999/12/a-third-way-for-the-third-world/377927/.

Lipset, Seymour Martin. "Some Social Requisites of Democracy: Economic Development and Political Legitimacy." *American Political Science Review* 53, no. 1 (1959): 69–105.

Marx, Karl. *Capital*. Library of Economics and Liberty. Accessed September 13, 2015, http://www.econlib.org/library/YPDBooks/Marx/mrxCpContents.html.

"The Measure of Progress." *Economist*, September 16, 1999. Accessed September 30, 2015. http://www.economist.com/node/239670?zid=295&ah=0bca374e65f2354d553956ea65f756e0.

Nguefack-Tsague, Georges, Stephan Klasen, and Walter Zucchini. "On Weighting the Components of the Human Development Index: A Statistical Justification." *Journal of Human Development and Capabilities* 12, no. 2 (2011): 183–202.

Nobelprize.org. "The Sveriges Riksbank Prize in Economic Sciences in Memory of Alfred Nobel 1998." Accessed September 30, 2015. http://www.nobelprize.org/nobel_prizes/economic-sciences/laureates/1998/.

Nussbaum, Martha. *Creating Capabilities: The Human Development Approach*. Cambridge, MA: Harvard University Press, 2011.

OCL Center. "WorldCat Identities: Amartya Sen 1933–." Accessed October 14, 2015. http://worldcat.org/identities/lccn-n50-12860.

Pollock, Robert L. "The Wrong Economist Won." *Wall Street Journal*, October 15, 1998. Accessed November 10, 2015. http://www.wsj.com/articles/SB908405798502442000.

Poverty.org. "Laeken Indicators." Accessed September 30, 2015. http://www.poverty.org.uk/summary/eu.htm.

Rawls, John. *A Theory of Justice*. Cambridge, MA: Harvard University Press, 1971.

Reid-Henry, Simon. "Amartya Sen: Economist, Philosopher, Human Development Doyen." *Guardian*, November 22, 2012. Accessed November 10, 2015. http://www.theguardian.com/global-development/2012/nov/22/amartya-sen-human-development-doyen.

Robbins, Lionel. *An Essay on the Nature and Significance of Economic Science*. London: Macmillan and Co., 1932.

Rodrick, Dani. "Goodbye Washington Consensus, Hello Washington Confusion? A Review of the World Bank's Economic Growth in the 1990s: Learning from a Decade of Reform." *Journal of Economic Literature* 44, no. 4 (2006): 973.

Ruger, Jennifer Prah. "Toward a Theory of a Right to Health: Capability and Incompletely Theorized Agreements." *Yale Journal of the Law and Humanities* 18, no. 2 (2006): 3.

Sen, Amartya. "Amartya Sen—Biographical." Nobelprize.org. Accessed October 14, 2015. http://www.nobelprize.org/nobel_prizes/economic-sciences/laureates/1998/sen-bio.html.

Collective Choice and Social Welfare. San Francisco: Holden Day, 1970.

Development as Freedom. New York: Anchor Books, 1999.

The Idea of Justice. Cambridge, MA: Harvard University Press, 2009.

Identity and Violence: The Illusion of Destiny. London: Penguin Books, 2006.

"More Than One Million Women Are Missing." *New York Review of Books*, 37, no. 20 (1990).

Poverty and Famines: An Essay on Entitlement and Deprivation. Oxford: Oxford University Press, 1981.

"Response to Commentaries." *Studies in Comparative International Development* 37, no. 2 (2002): 78–86.

Sen, Amartya, Bina Agarwal, Jane Humphries, and Ingrid Robeyns. "Continuing the Conversation." *Feminist Economics* 9 (2003): 319–32.

Sen, Amartya, and Martha Nussbaum, eds. *The Quality of Life*. New York: Oxford University Press, 1993.

Sinha, R. Sen, and Teresa Nobels. "A Christian Plot: Singhal." *The Indian Express*, December 28, 1998.

Smith, Adam. *An Inquiry into the Nature and Causes of the Wealth of Nations*. Library of Economics and Liberty. Accessed September 13, 2015. http://www.econlib.org/library/Smith/smWN1.html.

Smith, Geri. "Do Literacy and Health Spark Growth?" *Business Week*, September 20, 1999. Accessed September 30, 2015. http://www.businessweek.com/1999/99_38/b3647034.htm.

Srinivasan, Thirukodikaval Nilakanta. "Human Development: A New Paradigm or Reinvention of the Wheel?" *The American Economic Review* 84, no. 2 (1994): 238–43.

Stewart, Frances. "Women and Human Development: *The Capabilities Approach*, by Martha Nussbaum." *Journal of International Development* 13, no. 8 (2001): 1191–92.

"Groups and Capabilities." *Journal of Human Development* 6, no. 2 (2005): 185–204.

Stewart, Frances, and Séverine Deneulin. "Amartya Sen's Contribution to Development Thinking." *Studies in Comparative International Development* 37, no. 2 (2002): 61–70.

Stiglitz, Joseph E. *More Instruments and Broader Goals: Moving toward the Post-Washington Consensus*. Helsinki: UNU/WIDER, 1998.

Streeten, Paul, Shahid Javed Burki, Ul Haq, Norman Hicks, and Frances Stewart. *First Things First: Meeting Basic Human Needs in Developing Countries*. New York: Oxford University Press, 1981.

United Nations Development Programme. "2014 Human Development Index. Sustaining Human Progress: Reducing Vulnerabilities and Building Resilience." Accessed September 30, 2015. http://hdr.undp.org/en/content/table-1-human-development-index-and-its-components.

De Waal, Alex. "Famine Mortality: A Case Study of Darfur, Sudan 1984–85." *Population Studies* 43, no. 1 (1989): 5–24.

Williamson, John. "A Short History of the Washington Consensus." Paper commissioned by Fundación CIDOB for a conference "From the Washington Consensus towards a new Global Governance," Barcelona, September 24–25, 2004.

Zakaria, Fareed. "Beyond Money." *New York Times,* November 28, 1999. Accessed September 30, 2015. https://www.nytimes.com/books/99/11/28/reviews/991128.28zakarit.html.

THE MACAT LIBRARY BY DISCIPLINE

AFRICANA STUDIES

Chinua Achebe's *An Image of Africa: Racism in Conrad's Heart of Darkness*
W. E. B. Du Bois's *The Souls of Black Folk*
Zora Neale Huston's *Characteristics of Negro Expression*
Martin Luther King Jr's *Why We Can't Wait*
Toni Morrison's *Playing in the Dark: Whiteness in the American Literary Imagination*

ANTHROPOLOGY

Arjun Appadurai's *Modernity at Large: Cultural Dimensions of Globalisation*
Philippe Ariès's *Centuries of Childhood*
Franz Boas's *Race, Language and Culture*
Kim Chan & Renée Mauborgne's *Blue Ocean Strategy*
Jared Diamond's *Guns, Germs & Steel: the Fate of Human Societies*
Jared Diamond's *Collapse: How Societies Choose to Fail or Survive*
E. E. Evans-Pritchard's *Witchcraft, Oracles and Magic Among the Azande*
James Ferguson's *The Anti-Politics Machine*
Clifford Geertz's *The Interpretation of Cultures*
David Graeber's *Debt: the First 5000 Years*
Karen Ho's *Liquidated: An Ethnography of Wall Street*
Geert Hofstede's *Culture's Consequences: Comparing Values, Behaviors, Institutes and Organizations across Nations*
Claude Lévi-Strauss's *Structural Anthropology*
Jay Macleod's *Ain't No Makin' It: Aspirations and Attainment in a Low-Income Neighborhood*
Saba Mahmood's *The Politics of Piety: The Islamic Revival and the Feminist Subject*
Marcel Mauss's *The Gift*

BUSINESS

Jean Lave & Etienne Wenger's *Situated Learning*
Theodore Levitt's *Marketing Myopia*
Burton G. Malkiel's *A Random Walk Down Wall Street*
Douglas McGregor's *The Human Side of Enterprise*
Michael Porter's *Competitive Strategy: Creating and Sustaining Superior Performance*
John Kotter's *Leading Change*
C. K. Prahalad & Gary Hamel's *The Core Competence of the Corporation*

CRIMINOLOGY

Michelle Alexander's *The New Jim Crow: Mass Incarceration in the Age of Colorblindness*
Michael R. Gottfredson & Travis Hirschi's *A General Theory of Crime*
Richard Herrnstein & Charles A. Murray's *The Bell Curve: Intelligence and Class Structure in American Life*
Elizabeth Loftus's *Eyewitness Testimony*
Jay Macleod's *Ain't No Makin' It: Aspirations and Attainment in a Low-Income Neighborhood*
Philip Zimbardo's *The Lucifer Effect*

ECONOMICS

Janet Abu-Lughod's *Before European Hegemony*
Ha-Joon Chang's *Kicking Away the Ladder*
David Brion Davis's *The Problem of Slavery in the Age of Revolution*
Milton Friedman's *The Role of Monetary Policy*
Milton Friedman's *Capitalism and Freedom*
David Graeber's *Debt: the First 5000 Years*
Friedrich Hayek's *The Road to Serfdom*
Karen Ho's *Liquidated: An Ethnography of Wall Street*

John Maynard Keynes's *The General Theory of Employment, Interest and Money*
Charles P. Kindleberger's *Manias, Panics and Crashes*
Robert Lucas's *Why Doesn't Capital Flow from Rich to Poor Countries?*
Burton G. Malkiel's *A Random Walk Down Wall Street*
Thomas Robert Malthus's *An Essay on the Principle of Population*
Karl Marx's *Capital*
Thomas Piketty's *Capital in the Twenty-First Century*
Amartya Sen's *Development as Freedom*
Adam Smith's *The Wealth of Nations*
Nassim Nicholas Taleb's *The Black Swan: The Impact of the Highly Improbable*
Amos Tversky's & Daniel Kahneman's *Judgment under Uncertainty: Heuristics and Biases*
Mahbub Ul Haq's *Reflections on Human Development*
Max Weber's *The Protestant Ethic and the Spirit of Capitalism*

FEMINISM AND GENDER STUDIES

Judith Butler's *Gender Trouble*
Simone De Beauvoir's *The Second Sex*
Michel Foucault's *History of Sexuality*
Betty Friedan's *The Feminine Mystique*
Saba Mahmood's *The Politics of Piety: The Islamic Revival and the Feminist Subject*
Joan Wallach Scott's *Gender and the Politics of History*
Mary Wollstonecraft's *A Vindication of the Rights of Woman*
Virginia Woolf's *A Room of One's Own*

GEOGRAPHY

The Brundtland Report's *Our Common Future*
Rachel Carson's *Silent Spring*
Charles Darwin's *On the Origin of Species*
James Ferguson's *The Anti-Politics Machine*
Jane Jacobs's *The Death and Life of Great American Cities*
James Lovelock's *Gaia: A New Look at Life on Earth*
Amartya Sen's *Development as Freedom*
Mathis Wackernagel & William Rees's *Our Ecological Footprint*

HISTORY

Janet Abu-Lughod's *Before European Hegemony*
Benedict Anderson's *Imagined Communities*
Bernard Bailyn's *The Ideological Origins of the American Revolution*
Hanna Batatu's *The Old Social Classes And The Revolutionary Movements Of Iraq*
Christopher Browning's *Ordinary Men: Reserve Police Batallion 101 and the Final Solution in Poland*
Edmund Burke's *Reflections on the Revolution in France*
William Cronon's *Nature's Metropolis: Chicago And The Great West*
Alfred W. Crosby's *The Columbian Exchange*
Hamid Dabashi's *Iran: A People Interrupted*
David Brion Davis's *The Problem of Slavery in the Age of Revolution*
Nathalie Zemon Davis's *The Return of Martin Guerre*
Jared Diamond's *Guns, Germs & Steel: the Fate of Human Societies*
Frank Dikotter's *Mao's Great Famine*
John W Dower's *War Without Mercy: Race And Power In The Pacific War*
W. E. B. Du Bois's *The Souls of Black Folk*
Richard J. Evans's *In Defence of History*
Lucien Febvre's *The Problem of Unbelief in the 16th Century*
Sheila Fitzpatrick's *Everyday Stalinism*

Eric Foner's *Reconstruction: America's Unfinished Revolution, 1863-1877*
Michel Foucault's *Discipline and Punish*
Michel Foucault's *History of Sexuality*
Francis Fukuyama's *The End of History and the Last Man*
John Lewis Gaddis's *We Now Know: Rethinking Cold War History*
Ernest Gellner's *Nations and Nationalism*
Eugene Genovese's *Roll, Jordan, Roll: The World the Slaves Made*
Carlo Ginzburg's *The Night Battles*
Daniel Goldhagen's *Hitler's Willing Executioners*
Jack Goldstone's *Revolution and Rebellion in the Early Modern World*
Antonio Gramsci's *The Prison Notebooks*
Alexander Hamilton, John Jay & James Madison's *The Federalist Papers*
Christopher Hill's *The World Turned Upside Down*
Carole Hillenbrand's *The Crusades: Islamic Perspectives*
Thomas Hobbes's *Leviathan*
Eric Hobsbawm's *The Age Of Revolution*
John A. Hobson's *Imperialism: A Study*
Albert Hourani's *History of the Arab Peoples*
Samuel P. Huntington's *The Clash of Civilizations and the Remaking of World Order*
C. L. R. James's *The Black Jacobins*
Tony Judt's *Postwar: A History of Europe Since 1945*
Ernst Kantorowicz's *The King's Two Bodies: A Study in Medieval Political Theology*
Paul Kennedy's *The Rise and Fall of the Great Powers*
Ian Kershaw's *The "Hitler Myth": Image and Reality in the Third Reich*
John Maynard Keynes's *The General Theory of Employment, Interest and Money*
Charles P. Kindleberger's *Manias, Panics and Crashes*
Martin Luther King Jr's *Why We Can't Wait*
Henry Kissinger's *World Order: Reflections on the Character of Nations and the Course of History*
Thomas Kuhn's *The Structure of Scientific Revolutions*
Georges Lefebvre's *The Coming of the French Revolution*
John Locke's *Two Treatises of Government*
Niccolò Machiavelli's *The Prince*
Thomas Robert Malthus's *An Essay on the Principle of Population*
Mahmood Mamdani's *Citizen and Subject: Contemporary Africa And The Legacy Of Late Colonialism*
Karl Marx's *Capital*
Stanley Milgram's *Obedience to Authority*
John Stuart Mill's *On Liberty*
Thomas Paine's *Common Sense*
Thomas Paine's *Rights of Man*
Geoffrey Parker's *Global Crisis: War, Climate Change and Catastrophe in the Seventeenth Century*
Jonathan Riley-Smith's *The First Crusade and the Idea of Crusading*
Jean-Jacques Rousseau's *The Social Contract*
Joan Wallach Scott's *Gender and the Politics of History*
Theda Skocpol's *States and Social Revolutions*
Adam Smith's *The Wealth of Nations*
Timothy Snyder's *Bloodlands: Europe Between Hitler and Stalin*
Sun Tzu's *The Art of War*
Keith Thomas's *Religion and the Decline of Magic*
Thucydides's *The History of the Peloponnesian War*
Frederick Jackson Turner's *The Significance of the Frontier in American History*
Odd Arne Westad's *The Global Cold War: Third World Interventions And The Making Of Our Times*

The Macat Library By Discipline

LITERATURE

Chinua Achebe's *An Image of Africa: Racism in Conrad's Heart of Darkness*
Roland Barthes's *Mythologies*
Homi K. Bhabha's *The Location of Culture*
Judith Butler's *Gender Trouble*
Simone De Beauvoir's *The Second Sex*
Ferdinand De Saussure's *Course in General Linguistics*
T. S. Eliot's *The Sacred Wood: Essays on Poetry and Criticism*
Zora Neale Huston's *Characteristics of Negro Expression*
Toni Morrison's *Playing in the Dark: Whiteness in the American Literary Imagination*
Edward Said's *Orientalism*
Gayatri Chakravorty Spivak's *Can the Subaltern Speak?*
Mary Wollstonecraft's *A Vindication of the Rights of Women*
Virginia Woolf's *A Room of One's Own*

PHILOSOPHY

Elizabeth Anscombe's *Modern Moral Philosophy*
Hannah Arendt's *The Human Condition*
Aristotle's *Metaphysics*
Aristotle's *Nicomachean Ethics*
Edmund Gettier's *Is Justified True Belief Knowledge?*
Georg Wilhelm Friedrich Hegel's *Phenomenology of Spirit*
David Hume's *Dialogues Concerning Natural Religion*
David Hume's *The Enquiry for Human Understanding*
Immanuel Kant's *Religion within the Boundaries of Mere Reason*
Immanuel Kant's *Critique of Pure Reason*
Søren Kierkegaard's *The Sickness Unto Death*
Søren Kierkegaard's *Fear and Trembling*
C. S. Lewis's *The Abolition of Man*
Alasdair MacIntyre's *After Virtue*
Marcus Aurelius's *Meditations*
Friedrich Nietzsche's *On the Genealogy of Morality*
Friedrich Nietzsche's *Beyond Good and Evil*
Plato's *Republic*
Plato's *Symposium*
Jean-Jacques Rousseau's *The Social Contract*
Gilbert Ryle's *The Concept of Mind*
Baruch Spinoza's *Ethics*
Sun Tzu's *The Art of War*
Ludwig Wittgenstein's *Philosophical Investigations*

POLITICS

Benedict Anderson's *Imagined Communities*
Aristotle's *Politics*
Bernard Bailyn's *The Ideological Origins of the American Revolution*
Edmund Burke's *Reflections on the Revolution in France*
John C. Calhoun's *A Disquisition on Government*
Ha-Joon Chang's *Kicking Away the Ladder*
Hamid Dabashi's *Iran: A People Interrupted*
Hamid Dabashi's *Theology of Discontent: The Ideological Foundation of the Islamic Revolution in Iran*
Robert Dahl's *Democracy and its Critics*
Robert Dahl's *Who Governs?*
David Brion Davis's *The Problem of Slavery in the Age of Revolution*

Alexis De Tocqueville's *Democracy in America*
James Ferguson's *The Anti-Politics Machine*
Frank Dikotter's *Mao's Great Famine*
Sheila Fitzpatrick's *Everyday Stalinism*
Eric Foner's *Reconstruction: America's Unfinished Revolution, 1863-1877*
Milton Friedman's *Capitalism and Freedom*
Francis Fukuyama's *The End of History and the Last Man*
John Lewis Gaddis's *We Now Know: Rethinking Cold War History*
Ernest Gellner's *Nations and Nationalism*
David Graeber's *Debt: the First 5000 Years*
Antonio Gramsci's *The Prison Notebooks*
Alexander Hamilton, John Jay & James Madison's *The Federalist Papers*
Friedrich Hayek's *The Road to Serfdom*
Christopher Hill's *The World Turned Upside Down*
Thomas Hobbes's *Leviathan*
John A. Hobson's *Imperialism: A Study*
Samuel P. Huntington's *The Clash of Civilizations and the Remaking of World Order*
Tony Judt's *Postwar: A History of Europe Since 1945*
David C. Kang's *China Rising: Peace, Power and Order in East Asia*
Paul Kennedy's *The Rise and Fall of Great Powers*
Robert Keohane's *After Hegemony*
Martin Luther King Jr.'s *Why We Can't Wait*
Henry Kissinger's *World Order: Reflections on the Character of Nations and the Course of History*
John Locke's *Two Treatises of Government*
Niccolò Machiavelli's *The Prince*
Thomas Robert Malthus's *An Essay on the Principle of Population*
Mahmood Mamdani's *Citizen and Subject: Contemporary Africa And The Legacy Of Late Colonialism*
Karl Marx's *Capital*
John Stuart Mill's *On Liberty*
John Stuart Mill's *Utilitarianism*
Hans Morgenthau's *Politics Among Nations*
Thomas Paine's *Common Sense*
Thomas Paine's *Rights of Man*
Thomas Piketty's *Capital in the Twenty-First Century*
Robert D. Putman's *Bowling Alone*
John Rawls's *Theory of Justice*
Jean-Jacques Rousseau's *The Social Contract*
Theda Skocpol's *States and Social Revolutions*
Adam Smith's *The Wealth of Nations*
Sun Tzu's *The Art of War*
Henry David Thoreau's *Civil Disobedience*
Thucydides's *The History of the Peloponnesian War*
Kenneth Waltz's *Theory of International Politics*
Max Weber's *Politics as a Vocation*
Odd Arne Westad's *The Global Cold War: Third World Interventions And The Making Of Our Times*

POSTCOLONIAL STUDIES

Roland Barthes's *Mythologies*
Frantz Fanon's *Black Skin, White Masks*
Homi K. Bhabha's *The Location of Culture*
Gustavo Gutiérrez's *A Theology of Liberation*
Edward Said's *Orientalism*
Gayatri Chakravorty Spivak's *Can the Subaltern Speak?*

PSYCHOLOGY

Gordon Allport's *The Nature of Prejudice*
Alan Baddeley & Graham Hitch's *Aggression: A Social Learning Analysis*
Albert Bandura's *Aggression: A Social Learning Analysis*
Leon Festinger's *A Theory of Cognitive Dissonance*
Sigmund Freud's *The Interpretation of Dreams*
Betty Friedan's *The Feminine Mystique*
Michael R. Gottfredson & Travis Hirschi's *A General Theory of Crime*
Eric Hoffer's *The True Believer: Thoughts on the Nature of Mass Movements*
William James's *Principles of Psychology*
Elizabeth Loftus's *Eyewitness Testimony*
A. H. Maslow's *A Theory of Human Motivation*
Stanley Milgram's *Obedience to Authority*
Steven Pinker's *The Better Angels of Our Nature*
Oliver Sacks's *The Man Who Mistook His Wife For a Hat*
Richard Thaler & Cass Sunstein's *Nudge: Improving Decisions About Health, Wealth and Happiness*
Amos Tversky's *Judgment under Uncertainty: Heuristics and Biases*
Philip Zimbardo's *The Lucifer Effect*

SCIENCE

Rachel Carson's *Silent Spring*
William Cronon's *Nature's Metropolis: Chicago And The Great West*
Alfred W. Crosby's *The Columbian Exchange*
Charles Darwin's *On the Origin of Species*
Richard Dawkin's *The Selfish Gene*
Thomas Kuhn's *The Structure of Scientific Revolutions*
Geoffrey Parker's *Global Crisis: War, Climate Change and Catastrophe in the Seventeenth Century*
Mathis Wackernagel & William Rees's *Our Ecological Footprint*

SOCIOLOGY

Michelle Alexander's *The New Jim Crow: Mass Incarceration in the Age of Colorblindness*
Gordon Allport's *The Nature of Prejudice*
Albert Bandura's *Aggression: A Social Learning Analysis*
Hanna Batatu's *The Old Social Classes And The Revolutionary Movements Of Iraq*
Ha-Joon Chang's *Kicking Away the Ladder*
W. E. B. Du Bois's *The Souls of Black Folk*
Émile Durkheim's *On Suicide*
Frantz Fanon's *Black Skin, White Masks*
Frantz Fanon's *The Wretched of the Earth*
Eric Foner's *Reconstruction: America's Unfinished Revolution, 1863-1877*
Eugene Genovese's *Roll, Jordan, Roll: The World the Slaves Made*
Jack Goldstone's *Revolution and Rebellion in the Early Modern World*
Antonio Gramsci's *The Prison Notebooks*
Richard Herrnstein & Charles A Murray's *The Bell Curve: Intelligence and Class Structure in American Life*
Eric Hoffer's *The True Believer: Thoughts on the Nature of Mass Movements*
Jane Jacobs's *The Death and Life of Great American Cities*
Robert Lucas's *Why Doesn't Capital Flow from Rich to Poor Countries?*
Jay Macleod's *Ain't No Makin' It: Aspirations and Attainment in a Low Income Neighborhood*
Elaine May's *Homeward Bound: American Families in the Cold War Era*
Douglas McGregor's *The Human Side of Enterprise*
C. Wright Mills's *The Sociological Imagination*

Thomas Piketty's *Capital in the Twenty-First Century*
Robert D. Putman's *Bowling Alone*
David Riesman's *The Lonely Crowd: A Study of the Changing American Character*
Edward Said's *Orientalism*
Joan Wallach Scott's *Gender and the Politics of History*
Theda Skocpol's *States and Social Revolutions*
Max Weber's *The Protestant Ethic and the Spirit of Capitalism*

THEOLOGY

Augustine's *Confessions*
Benedict's *Rule of St Benedict*
Gustavo Gutiérrez's *A Theology of Liberation*
Carole Hillenbrand's *The Crusades: Islamic Perspectives*
David Hume's *Dialogues Concerning Natural Religion*
Immanuel Kant's *Religion within the Boundaries of Mere Reason*
Ernst Kantorowicz's *The King's Two Bodies: A Study in Medieval Political Theology*
Søren Kierkegaard's *The Sickness Unto Death*
C. S. Lewis's *The Abolition of Man*
Saba Mahmood's *The Politics of Piety: The Islamic Revival and the Feminist Subject*
Baruch Spinoza's *Ethics*
Keith Thomas's *Religion and the Decline of Magic*

COMING SOON

Chris Argyris's *The Individual and the Organisation*
Seyla Benhabib's *The Rights of Others*
Walter Benjamin's *The Work Of Art in the Age of Mechanical Reproduction*
John Berger's *Ways of Seeing*
Pierre Bourdieu's *Outline of a Theory of Practice*
Mary Douglas's *Purity and Danger*
Roland Dworkin's *Taking Rights Seriously*
James G. March's *Exploration and Exploitation in Organisational Learning*
Ikujiro Nonaka's *A Dynamic Theory of Organizational Knowledge Creation*
Griselda Pollock's *Vision and Difference*
Amartya Sen's *Inequality Re-Examined*
Susan Sontag's *On Photography*
Yasser Tabbaa's *The Transformation of Islamic Art*
Ludwig von Mises's *Theory of Money and Credit*

Printed in the United States
by Baker & Taylor Publisher Services